Advances in Circuits and Systems

SELECTED PAPERS ON

LOGIC SYNTHESIS FOR INTEGRATED CIRCUIT DESIGN

EDITED BY

A. RICHARD NEWTON

**Department of Electrical Engineering and
Computer Sciences
University of California, Berkeley**

*A SERIES PUBLISHED FOR THE
IEEE CIRCUITS AND SYSTEMS SOCIETY*

IEEE
PRESS

The Institute of Electrical and Electronics Engineers, Inc., New York

Copyright © 1987 by
THE INSTITUTE OF ELECTRICAL AND ELECTRONICS ENGINEERS, INC.
345 East 47th Street, New York, NY 10017-2394
All rights reserved.

PRINTED IN THE UNITED STATES OF AMERICA

IEEE Order Number: PP0226-1

Library of Congress Cataloging-in-Publication Data

Selected papers on logic synthesis for integrated circuit design.

(Advances in circuits and systems)
''A series published for the IEEE Circuits and Systems Society.''
Includes index.
1. Integrated circuits—Very large scale integration. 2. Logic design. I.
Newton, A. Richard (Arthur Richard), 1951–   . II. IEEE Circuits and
Systems Society. III. Series.
TK7874.S414   1987   621.381'73   87-7369
ISBN 0-87942-236-X

Contents

Introduction

Computer aids have been used for both the design and verification of electronic systems for many years prior to the introduction of commercial Integrated Circuits (ICs) in the early 1960s. Such tools have found their way into virtually every aspect of the design of such systems, from integrated circuit process technology to the design of complex computer architectures. The recent explosion in the complexity of electronic systems that the advent of Very Large Scale Integration (VLSI) has allowed, has made the use of sophisticated computer-aided design tools indispensable. Not only are computer aids necessary for both the design and verification of integrated circuits today but, as the semiconductor processing technologies mature, computer aids will soon also provide key proprietary advantages as semiconductor and system design houses vie for the promising Application-Specific IC (ASIC) market of the next decade. The pivotal technologies in future IC CAD systems include *tools for logic synthesis*, such as combinational and sequential logic synthesis tools, and architectural design aids; *tools for physical design*, including placement, routing, and technology re-mapping tools; *design system management tools*, including the management of design versions and alternatives in a distributed computing environment, data dependency management, and efficient and flexible interfaces to new tools; *verification tools*, including physical and electrical rules checking, simulation, and formal verification techniques.

Because of the decrease in the relative competitive advantage obtained from process technology, semiconductor companies and "silicon foundries" must emphasize other aspects of the design process if they are to compete effectively for the ASIC market. The two avenues available are in architecture — hiring better designers and system architects than their competitors, which is often difficult and is certainly expensive — and in computer-aided design. A lack of CAD support in any of the above areas may result in a significant reduction in the competitiveness of the designers' final product. On the other hand, a significant proprietary advantage in any or all of these areas will maintain a company's position as a force in the marketplace.

Over the past decade, significant progress has been made in reducing the time required for the physical design of high-performance integrated circuits and systems. However, with a few notable exceptions, very little progress has been made in improving the system, register-level, and logic gate-level design time for high-quality results. With the physical design problem no longer dominating the design cycle, the automated design of two-level and multi-level combinational logic has begun to receive increased attention. Over the next few years, the areas of sequential machine design, architectural trade-off analysis, and synthesis of final designs from behavioral specifications promise to be particularly fertile areas for research and have the potential to provide a high industrial pay-back.

Much of the early work in the area of logic synthesis began in the 1950's, particularly at the computer companies, and most notably at IBM Corporation. However, only in the past decade has a significant amount of work been directed towards applying logic synthesis to integrated-circuit-specific problems, such as the implementation of large, Programmable Logic Array structures or the optimal use of Standard Cell libraries. Most of this work has been directed towards area optimization of such implementations and only recently have technology-specific timing models been included in the optimization to meet specific system timing requirements. Indeed, it is in this important area that a significant amount of work remains.

It is not possible to include all of the historically significant papers in this field here. Nor is it possible to do justice to the breadth of work involved in this field. Rather, I have selected a number of more recent papers, which reference most of the classic works on which they have built, yet which point out the issues specific to integrated circuit and system implementations of digital designs.

March 12, 1987

A. Richard Newton
University of California
Berkeley, California.

1

CAD Tools for ASIC Design

ARTHUR R. NEWTON, MEMBER, IEEE, AND ALBERTO L. SANGIOVANNI-VINCENTELLI, FELLOW, IEEE

Invited Paper

Computer aids have been used for both the design and verification of electronic systems for many years. The recent explosion in the complexity of electronic systems that the advent of Very Large Scale Integration (VLSI) has allowed, has made the use of sophisticated computer-aided design tools indispensable. Computer aids will soon also provide key proprietary advantages as semiconductor and system design houses vie for the promising Application-Specific IC (ASIC) market of the next decade. This paper focusses on the techniques critical to both custom and ASIC design, the directions of present research and development for these areas, and future trends. In particular, recent developments in tools for the automated design of combinational logic are reviewed. These techniques include both algorithmic and rule-based approaches.

I. INTRODUCTION

Computer aids have been used for both the design and verification of electronic systems for many years prior to the introduction of commercial Integrated Circuits (ICs) in the early 1960s. Such tools have found their way into virtually every aspect of the design of such systems, from integrated circuit process technology to the design of complex computer architectures. The recent explosion in the complexity of electronic systems that the advent of Very Large Scale Integration (VLSI) has allowed, has made the use of sophisticated computer-aided design tools indispensable. Not only are computer aids necessary for both the design and verification of integrated circuits today but, as the semiconductor processing technologies mature, computer aids will soon also provide key proprietary advantages as semiconductor and system design houses vie for the promising Application-Specific IC (ASIC) market of the next decade. Computer aids are also playing an increasing part in the design and verification of commodity circuits, including complex microcomputers and memory devices. We believe that the pivotal technologies in future IC CAD systems include the following: *tools for IC synthesis*, such as place-ment and routing, combinational and sequential logic synthesis tools, and architectural design aids; *design system management tools*, including the management of design versions and alternatives in a distributed computing environment, data dependency management, and efficient and flexible interfaces to new tools; and *verification tools*, including physical and electrical rules checking, simulation, and formal verification techniques.

Nowadays, the field of CAD for IC design is very broad and it is not possible to cover all aspects of CAD for VLSI in a single paper. For this reason, the paper is focussed on the techniques critical to both custom and ASIC design, the directions of present research and development for these areas, and future trends. While the areas of procedural design, "silicon compilation," design verification, and testing are of great importance, they are dealt with in other papers in this issue and for that reason they are only mentioned briefly here.

For an ASIC design environment, a simplified designer's view of the role of CAD is illustrated in Fig. 1. The designer,

Manuscript received August 15, 1986; revised December 8, 1986. This research has been sponsored in part by SRC, DARPA, the MICRO program of the state of California, and our industrial affiliates.

The authors are with the Department of Electrical Engineering and Computer Sciences, University of California, Berkeley, CA 94720, USA.

IEEE Log Number 8714508.

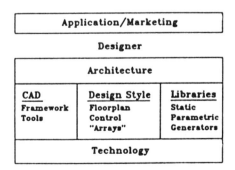

Fig. 1. Major components of ASIC design flow.

driven by the marketing need for a circuit that meets a particular cost, performance, and functional specification, works with system, logic, and circuit architectures to create a chip design. In that process, the CAD tools are used to evaluate tradeoffs and alternative designs, to construct specific circuit components, and to assemble and interconnect the components to form the final chip patterns. Once the IC mask patterns have been assembled, CAD tools are used

Reprinted from *Proc. IEEE*, vol. 75, no. 6, pp. 765–776, June 1987.

3

to check the final layout and prepare it for the automated manufacture of masks. As the competition for designs increases, together with the increasing number of companies in the ASIC business and by the high capital cost of a modern IC processing facility, there is increasing demand for designers to be able to differentiate their IC product from that of their competitors. Higher performance, lower cost, more features, and a faster time to market are all major factors which differentiate IC products in the ASIC marketplace. In the past, different companies have been able to provide such product differentiation through their IC fabrication technologies: the ability to pack more transistors on a given chip or to provide a higher switching speed per gate drove the designs and their advantage in the marketplace. High-quality products as well as excellent pre- and post-sales support also play important roles in the industry today and will remain as important factors. However, the silicon planar process technology is maturing rapidly—significant gains in the performance and density are becoming increasingly expensive and many companies are resorting to "joint ventures," often with former competitors in the United States, Europe, or Asia, to maintain their position in IC process technologies. Because of this decrease in the relative competitive advantage obtained from process technology, semiconductor companies and "silicon foundries" must emphasize other aspects of the design process if they are to compete effectively for the ASIC market. The two avenues available are in architecture—hiring better designers and system architects than their competitors, which is often difficult and is certainly expensive—and in computer-aided design.

As illustrated in Fig. 1, the design task involves three major components: CAD programs, support for specific design styles, and support for component libraries. A lack of CAD support in any of these areas may result in a significant reduction in the competitiveness of the designers' final product. On the other hand, a significant proprietary advantage in any or all of these areas will maintain a company's position as a force in the marketplace.

The CAD tools area may be subdivided further into three areas: tools for circuit design, or synthesis; tools for verification of the design; and tools for design data management and for managing the flow of the design process. This last category of tools is of particular importance for it provides the foundation on which the CAD system is built. If the design *framework* is inflexible and cannot adapt to new tools, new design styles, and changes in process technology, then the design system will soon become obsolete. It is also important in maintaining a competitive advantage in the design process since an *open* framework, which supports the addition of locally developed as well as commercial tools, can be used to provide a proprietary difference between the competing systems.

Since design verification has received a great deal of attention in the past [1], most of the techniques and tools are relatively mature and are not reviewed here. However, we would like to point out that the major research issues in the verification area concern improving the performance of the tools for large designs without sacrificing reliability of the results. The use of special-purpose hardware and new computer architecture are playing a major role in this area [2]. In addition, new algorithms are being developed which exploit the properties of large circuits, such as the repetitive

use of circuit structures. Many of the new techniques, while novel and requiring large engineering investments to achieve their potential, are relatively easy to duplicate and therefore cannot provide the foundation for a proprietary technology.

On the other hand, with a few notable exceptions, design synthesis systems for ASIC designs are far less mature and large gains in circuit efficiency and design time are still to be had. In addition, many of the state-of-the-art synthesis techniques involve far more "inspiration" than "perspiration" and, as a result, can form the basis of a proprietary and differentiating technology for ASIC design. Techiques for efficient synthesis (system design, register-level design, logic design, placement, routing, and array compilation) will provide a major focus for both University research and Industrial competition over the coming years. In this paper, the optimization and descriptive aspects of the synthesis process are presented, with particular emphasis on *combinational logic*. Large blocks of combinational logic are often needed to implement the control aspects of the circuit. It is the design of this logic which often takes a significant amount of time if the logic is to be optimized for high performance.

The second important area for differentiation is that of CAD support for design styles. In particular, CAD support for floorplan style (gate-array, standard cell, macrocell, etc.) and support for the design of so-called "random" logic— that portion of a design that cannot be cast into a straightforward and efficient regular layout style, such as RAM, ROM, or datapath. Since designers are finding improved circuit design styles and layout styles continuously, it is essential that a CAD system be able to support a variety of design styles and adapt easily to new developments in these areas.

Finally, all ASIC systems require a library of primitive components, whether they be individual transistors, logic gates, or entire subsystems. These library cells may be invariant designs, such as the traditional standard cell or gate-array building blocks, they may be parameterized cells, such as those in the libraries offered by the "silicon compiler companies," or they may be sophisticated, module-generator-based libraries, where different cell topologies are generated on the fly as a function of the user's input description.

If a designer is to compete in the competitive ASIC marketplace of tomorrow, he must be able to customize his CAD design environment in all three of these areas.

This paper is organized as follows. In Section II, an overview of the synthesis process as related to different *design styles* is presented. In Section III, the synthesis of PLA-based structures and multi-level synthesis techniques are described, and in Section IV, we offer concluding remarks.

II. DESIGN METHODS AND SYNTHESIS

A. Design Methods

The use of a particular class of circuit structures is referred to as a *design method*, or design style, and while the development of new algorithms and techniques for CAD continues, a significant contribution to the design of VLSI circuits will continue to come from the development of new circuit design methods. However, while the implementation of a design method does not *require* the use of computer aids *per se*, the most successful design methods will be those designed to take maximum advantage of the computer in both the circuit design and verification phases. The design method must provide the *structure* necessary to use both human and computer resources effectively. For VLSI, this structure also provides the reduction in design complexity necessary to reduce design time and to ensure that the circuit function can be verified and the resulting circuit can be tested. In describing the variety of computer aids used for IC design, a distinction is made between those techniques used for *design*, or synthesis, of the IC and those techniques used for its *verification*. In both of these categories, a further distinction is made between techniques relating to the *physical*, or topological, aspects of the design process, such as the generation and verification of mask layout data or the placement of components in a circuit, and *functional* considerations, such as logic description, synthesis, simulation, and test-pattern generation.

Computer aids for design, or synthesis, at both the functional and physical levels, are primarily concerned with the use of *optimization* to improve performance and cost. These design tasks may be formulated as combinatorial optimization problems for operations such as cell placement, routing, logic minimization, and logic state assignment, or as parametric optimization problems for operations such as design at the electrical level.

Design methods can be classified in four categories: programmable arrays, standard-cell, macrocell, and procedural design. A VLSI circuit may consist of one large building block or it may consist of a number of building blocks combined either manually or by a computer program.

A programmable array is a one- or two-dimensional array of repeated cells which can be customized by adding or deleting geometry from specific mask layers. Since a number of processing steps are completed prior to customization, the locations of components on those layers are independent of a particular circuit implementation. Examples of programmable arrays include the Gate-Array, Weinberger Array [3], Programmable Logic Array (PLA) [4], [5], and Read-Only Memory (ROM).

The *gate-array* (also referred to as master-slice, or uncommitted logic array) is by far the most common programmable array designed by computer. Consequently, the computer aids for gate-array design are also the most advanced and the most mature. In this approach, a two-dimensional array of replicated transistors is fabricated to a point just prior to the interconnection levels. A particular circuit function is then implemented by customizing the connections within each local group of transistors, to define its characteristics as a basic cell (this usually resides in a cell library), and by customizing the interconnections between cells in the array to define the overall circuit (placement and routing). Generally, a two-level interconnection scheme is used for signals and, in some approaches, a third, more coarsely defined layer of interconnections is provided for power and ground connections. Because one or more interconnection layers are used within a group of transistors to define the function of a cell, these intra-cell interconnections often block the passage of more global inter-cell connections. For that reason, and to simplify the placement and routing problems associated with these arrays, the inter-cell connections are implemented on a rectilinear grid in the *channels* between the cells. In some cases, channels are also provided which run over the cells themselves and in some arrays, wider channels are provided in the center of the array to alleviate the congestion often found in that area if particular routing strategies are employed.

As processing technology advances and additional levels of high-quality interconnect are provided, the need for predetermined wiring channels is less clear. The gate-array placement and routing problem becomes much closer to that of a printed circuit board, where routing is possible over all cells, and the individual transistors can be packed closely together. This recent variation of the gate-array structure is often referred to as a "sea-of-gates" array [6] and, as well as providing new challenges for the CAD community, promises to replace many of the designs previously undertaken in the conventional gate-array style. In addition, due to the flexibility of the approach, we expect that such layout styles will also be used in preference to standard cell and macrocell for many circuits.

PLAs may also be used to implement building blocks directly, with storage elements in the feedback path to implement sequential logic in the classical Moore or Mealy style [7]. The PLA consists of a number of transistor arrays which implement logic AND and OR operations. In MOS technology, NOR-NOR arrays are used [8]. A conventional PLA consists of two arrays of cells: an input, or lookup plane followed by an output plane. A *folded* PLA may use additional planes, since rows and/or columns in the structure may be shared by more than one circuit variable, as described later.

The *standard cell* (or polycell) approach refers to a design method where a library of custom-designed cells is used to implement a logic function. These cells are generally of the complexity of simple logic gates or flip-flops and may be restricted to constant height and/or width to aid packing and ease of power distribution. Nowadays, however, state-of-the-art standard cell systems permit cells of different height *and* width to be included in the same design. This results in nonuniform routing channels between adjacent rows and requires a more sophisticated channel routing capability if the silicon area is to be used to its maximum efficiency. Unlike the programmable array approach, standard cell layout involves the customization of all mask layers. This additional freedom permits variable-width channels to be used. While most standard cell systems only permit inter-cell wiring in the channels between rows of cells or through cells via predetermined "feedthrough" cells, some systems permit over-cell routing if additional levels of interconnect are available. Standard cell systems are also used extensively in a variety of technologies including bipolar and CMOS.

It is often relatively inefficient to implement all classes of logic functions in a single design approach. For example, a standard cell approach is inefficient for memory circuits such as RAM and stack. In the *macrocell*, or building block, method, large circuit blocks, customized to a certain type of logic function, are available in a circuit library. These blocks are of irregular size and shape and may allow functional customization via interconnect, such as a PLA or ROM macro, or they can be parameterized with respect to topology as well. With the parameterized cell, the number of inputs and outputs may be parameters of the cell. In some systems macro cells may also be embedded in gate-array or standard-cell designs. The macrocell floorplan style is evolving as the floorplan of choice for large, ASIC designs.

All of the design methods described above may be classified as *data-driven*. That is, a description of the required logic function, in the form of equations or an interconnection list, is used as input to a software system which interprets the data and generates the final design. For gate-arrays and standard cells this is often referred to as a *semicustom* design style. Techniques have been developed over the last few years which can be classified as *procedure*, or *program*, driven [9]. These *procedural design* approaches, as well as their advantages and limitations as implemented today, are described in another paper in this issue [10]. Most of the 'silicon compiler' companies of today support macrocell-based floorplans, with procedurally based *module generators*.

B. Synthesis of Integrated Circuits

A complete synthesis system should generate layout masks from a high-level *algorithmic*, *behavioral*, or *functional* description of a VLSI system, a description of the target technology, and a description of the design constraints and cost functions. The design should be completed in reasonable time and with the quality a human designer could obtain, or better.

Very few design aids are available to assist the VLSI designer at the algorithmic level. At this level, the designer describes the system by specifying its operations or functions without giving implementation details such as the "hardware" components needed to implement the system. A major problem which remains today at this level of design is the definition of a notation for capturing the behavior of a design without ambiguity. Most languages that have been developed for this level of specification have imprecise semantics, making them unacceptable for input to automatic synthesis systems. Design at the architectural level involves the translation of a required algorithmic-level specification into a *register-transfer-level* implementation. This representation of the design includes components such as registers, memories and processors, which implement the high-level specification of the system.

Once the functional partitioning of the design is completed, estimates of the layout size, power-supply requirements, and speed of the high-level circuit blocks used to implement the various subfunctions are required. A chip plan must also be constructed to determine the relative placement of these building blocks. This chip plan is then further refined as the design proceeds. These tasks are often performed manually, perhaps with the help of the computer to perform book-keeping tasks such as the storage of interconnection data.

Silicon compilers have been proposed to carry out the entire synthesis process, as described in a companion paper [10]. The present trend in this area is to break the synthesis process into stages, and to use tools that optimize area and/or performance from one stage to the next. To differentiate this approach from the one taken by early silicon compilers, where the emphasis was simply on achieving a layout, we call systems constructed using these optimized tools *synthesis systems*.

The three basic components of a synthesis system are as follows:

1) Physical synthesis or layout, including floor planning, partitioning, placement, routing, and compaction.

2) Logic synthesis, including combinational logic, sequential logic, and algorithmic or behavioral synthesis.

3) Procedural design and module generation.

For lack of space, we concentrate on one aspect: the logic synthesis step, combinational logic, and sequential logic synthesis; while we direct the interested reader to other review papers on placement, routing, module generation, and behavior synthesis [1], [11]–[17].

III. LOGIC SYNTHESIS

A. Introduction

Over the past few years, placement and routing techniques have been developed which perform reasonably well for most block-oriented design styles. However, the synthesis of the circuit itself—deciding how to partition the logic, in what form to implement specific pieces of the logic, and what layout style to use for implementation—is still a largely manual process. For a processor circuit, the control logic portion of the chip is often the most time-consuming piece to design, is generally on the critical path for timing, and is often implemented in a very inefficient way. Automated synthesis of the control logic blocks of a chip, optimized for speed and area, provides one of the major challenges facing CAD today. In this section, the state of the art for the synthesis of combinational and sequential, two-level and multi-level synthesis is presented. Areas which provide the most potential for improvement are mentioned and recent work in this area is described.

B. PLA-Based Synthesis

Programmable Logic Arrays (PLAs) are perhaps the most popular structures for the implementation of two-level logic functions. Most modern VLSI microprocessors include large PLAs to implement the datapath control, as well as a variety of smaller PLAs for controlling other activities on the chip. Other chips, such as memory management circuits are often almost all PLAs. There are a number of reasons why PLAs are used so often today:

1) It is easy to implement a logic function in PLA form and there is a low probability of error. Once the PLA personality matrix has been obtained from a truth table or from Boolean logic equations, there is a one-to-one correspondence between the logic '1's and logic '0's of the personality matrix and with the cells used to build the PLA itself.

2) A computer program can be written to lay out PLAs automatically without requiring complex algorithms or large amounts of CPU time.

3) It is straightforward to make an 'Engineering Change' (EC) to a PLA at the last minute, without disturbing the surrounding circuitry.

4) Many tools are available for the minimization and layout of PLAs, as well as for test pattern generation.

Many PLA layout generators have been written based on simple translations of the Boolean equations into the layout of the PLA, e.g., [18], [19]. However, a straightforward implementation of the logic entered by the designer may result in PLAs which are large and, as a result, will have poor performance in terms of speed and power.

1) Combinational Logic: It is clear that PLA optimization

is necessary to obtain an effective implementation. The optimization steps involved in the transformation of combinational logic into the layout of a PLA are as follows:

a) Logic-level, or *functional*, optimization which aims at the reduction of the number of product terms needed to implement the function.

b) *Topological* optimization which aims at the elimination of unused space inside the core of the PLA, e.g., folding and simple partitioning.

c) Layout and circuit optimization, which attempts to perform optimal sizing and placement of drivers, loads, core cells, and additional ground lines.

In this section, we will consider only the first optimization step, while for lack of space, we direct the interested reader to [20], [22] for the other steps.

Over the past few years, a great deal of attention has been paid to logic minimization of two-level logic [23]–[25] (see [26] for other references). Recent research on logic-minimization algorithms produced a series of very efficient logic minimizers, ESPRESSOII, ESPRESSO-IIC, and ESPRESSO-MV [26], [29], [30]. ESPRESSO-IIC has been found to be very effective in minimizing complex logic functions while consuming a reasonable amount of computer resources. Other logic minimizers such as TAU [32] and the minimizer in PLASCO [33] used similar concepts to provide fast logic minimization. When the logic function is implemented using a PLA, logic mininization both reduces the area occupied by the PLA and improves its electrical performance. The algorithmic complexity of complete logic minimization is very high and so heuristic logic minimizers are used when medium and large logic functions have to be minimized.

However, simple logic minimization is not sufficient in most cases and other factors must be taken into account. Consider the case of two of the instruction-decode PLAs used in a recent chip designed at Berkeley, the SOAR (Smalltalk on an RISC) [34] microprocessor. The CMOS implementation of those PLAs, generated automatically by our PLA tool set at Berkeley, is shown in Fig. 2 [35]. Note that the same inputs are routed to each PLA (actually, one input is not needed in each PLA and so there are nine inputs to each one rather than the ten inputs used to describe the entire function). The input plane is at the bottom of each PLA. The SOAR designers decided to split the logic into two separate pieces since it appeared the PLA would otherwise be too big. In fact, when the logic equations were translated directly to a PLA personality matrix, with little optimization, there were 152 product terms for the entire decode function. After logic minimization using ESPRESSO-IIC [29], that number was reduced to 80 terms for the total PLA (T), still considered too large by the SOAR designers. So the logic was carefully partitioned into two pieces, PLA (A) and PLA (B), by separating the outputs into two groups of 16 and 23, respectively. After minimization, these PLAs required 38 and 41 product terms, respectively, as shown in Fig. 2. This was the "optimized" implementation used for the fabrication run.

In general, when a logic function is specified, only the 'on-set' is specified explicitly; i.e., when the function should take on the value logic-'1'. Since the 'off-set' is not specified explicitly, many PLA design systems make the assumption that all unspecified logic values are logic-'0'. Unfortunately, many of those states may, in fact, be 'don't-cares'; i.e., the output may be in either state logic-'0' or logic-'1'. For the case of the SOAR PLAs above, one of the input signals was RESET. When RESET was logic-'0' (an active-low signal), only four outputs from the control logic were of any consequence, the others could take on either state. By noting this fact, the 'don't-cares' were coded [30] into the output plane of the PLA personality matrix and allowed ESPRESSO-IIC to perform minimization [15] using these conditions. The result was a reduction in the number of product terms in PLA (T) from 80 to 63, or 21 percent. Of course, this was only one of many possible 'don't-care' conditions for this function. With further examination of the logic functions and an understanding of the operation of SOAR, further reductions would be possible. A better solution would have been

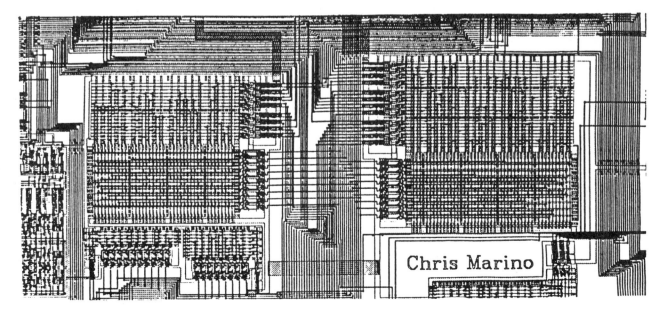

Fig. 2 The CPLAs from Berkeley CMOS SOAR.

to provide a logic specification system which would permit the designers to capture either the 'off-set' or the 'don't cares' explicitly.

Still further reduction can be achieved by taking care to assign the optimal output phase for the function. Since each output of the PLA is buffered anyway, the output buffer can be inverting or noninverting without affecting the performance or area of the design. By careful choice of whether to optimize for an output or its complement, further product-term sharing and product-term reduction can be achieved, at the same time reducing the number of transistors used in the PLA and therefore making the PLA more amenable to topological folding later. In [36], an approach was presented for the selection of phases which sometimes leads to a substantial reduction in the logic for a PLA. This scheme has been implemented in ESPRESSO-MV, and with optimal assignment of output phases, PLA (T) was reduced to 47 product terms—a total reduction of 41 percent from the original, minimized PLA. While this is an impressive result, further work is needed in this area. In fact, by requiring ESPRESSO-MV to implement the complement of *every* output variable, the number of product terms was reduced even further to 39 terms, or a total reduction of over 50 percent! Output phase assignment is, in fact, simply a special case of the general output encoding problem [16].

With all of these logic optimizations included, the resulting "total" PLA has as few product terms as each of the partitioned PLAs. Hence, the partitioning does, in fact, increase the overall area since the inputs and product-term loads must be duplicated for each PLA. This would still be the case after optimizing the separate pieces of the logic, as demonstrated by the last two columns of Table 1.

2) Finite-State Machines: With at least a resonable understanding of PLA-based combinational logic implementation, the next step was to attempt the design of PLA-based sequential circuits—in particular, the problem of generating optimal PLA-based Finite-State Machines (FSMs). Such circuits use a PLA to implement the combinational part of the logic, and the secondary outputs are fed back to the secondary inputs of the PLA via clocked latches. For a given set of primary (external) inputs and a required set of primary outputs, the objective is to:

a) choose the number of secondary outputs to be fed back, via the latches, as inputs and

b) assign values to these outputs (logic-'1' or logic-'0') for each state specified in the FSM description

such that the total area occupied by the combinational logic and/or the critical-path delay through the PLA are minimized. The 'textbook' approach to this problem, originated

in the days of discrete SSI circuits, is to choose the number of secondary variables so as to minimize the number of latches used. By doing so, one minimized the number of expensive IC packages needed for latches. To reduce the amount of combinatinal logic, various heuristic schemes were used. The most common approach was to use a 'distance-one' state encoding for adjacent states. Unfortunately, such a state-encoding strategy does not work well for PLA-based FSM designs.

To solve this problem optimally for PLA-based designs, an efficient approach to state assignment is needed. Many algorithms have been proposed in the past, e.g., [37], [38] to perform optimal state assignment. However, the results obtained were not satisfactory because of the complexity of the algorithms suggested or the performance of the PLA which implements the combinational part of the FSM. A new approach for performing an optimal assignment of binary codes to a set of symbolic inputs has been developed in [39]. The idea used in this approach is to view this assignment problem as a multiple-valued logic minimization problem. This multiple-valued minimization problem can be solved using a standard binary-valued minimization algorithm (e.g., ESPRESSO-IIC) or, preferably, it can be solved directly with a multiple-valued minimization algorithm, e.g., ESPRESSO-MV [31].

The program KISS [39] was developed to determine a state assignment based on encoding the symbolic present-state input variable of an FSM. For example, consider an industrial FSM with a 5-bit, minimal-length state vector (and three additional outputs). In a 3-μm p-well CMOS process, this circuit occupied 0.75 mm². After processing the FSM description using KISS, a 9-bit state vector was chosen. This resulted in four additional output columns, but reduced the overall area to 0.39 mm² by reducing the number of product terms substantially [15].

The KISS approach is successful, but an extension to this technique was needed which would allow for the complementary output-encoding problem, and which would eventually solve both problems simultaneously [16]. Recently, a series of algorithms have been developed for the optimal output encoding problem [21]. The techniques originated by these algorithms have been implemented in CAPPUCCINO [21], which tries to combine the optimal input encoding algorithms with the optimal output encoding ones.

Further work is needed to partition a large FSM into smaller machines where the intermediate output values are encoded. Though this process may result in more stages of logic, in many cases it is expected that the increase in clock speed that can be achieved using the small PLAs will more than compensate for the added clock cycles. Once again,

Table 1

PLA	Inputs	Outputs	Simple Minimization	With Output RESET 'don't cares' Specified	With Output Phase Assignment	With All Outputs Complemented
(A)	9	16	38	30	27	31
(B)	9	23	41	37	26	30
(T)	10	39	80	63	47	39

algorithms have been proposed in the past [41], but the technological constraints and objectives which drive the decomposition have changed drastically so as to make the existing algorithms inappropriate.

C. Multi-Level Logic Synthesis

As seen in the previous section, a great deal of work has been done to implement combinational logic in optimal, two-level form using the PLA. However, some control logic has a two-level representation which can have as many as 2^n product terms, where n is the number of primary inputs of the logic, even after minimization. In addition, even if a two-level representation contains a reasonable number of terms, there are cases in which a multi-level representation can be implemented in much less area and generally as a much faster circuit. In fact, a two-level logic representation can be viewed as a special case of general multi-level representations. Hence, a general framework for control logic design should offer multi-level synthesis tools which are able to select a two-level implementation wherever the two-level form is more effective in terms of area and/or speed. To be able to explore the design tradeoffs, such a system should offer a variety of both electrical design style (e.g., Domino logic, static CMOS) and layout design style (e.g., Weinberger Arrays, Gate Matrix, Standard Cells, and Gate Arrays) alternatives.

Multi-level logic synthesis is a very difficult problem and much work must be done to bring this area to the same level of advancement as two-level logic synthesis. However, in the past eight years, intense research activities have produced a variety of promising algorithms and effective synthesis systems. The first such system was the IBM Logic Synthesis System (LSS) [42], which has as target technology a variety of gate arrays and which has been recently extended to the so-called *open book* technologies, i.e., for standard cells and CMOS dynamic structures [43]. The Yorktown Silicon Compiler [27] has Cascode Voltage Switches [50] as its target technology. The SOCRATES system [33] has as target technologies gate arrays and standard cells. The recently developed MIS system [44] has as target technology CMOS static complex gates or macrocells, but as the YSC, its algorithms can easily support a variety of target technologies.

The optimization criterion of all multi-level logic synthesis systems is to minimize the area occupied by the logic equations (which is measured as a function of the number of gates, transistors, and nets in the final set of equations) while simultaneously satisfying the timing constraints derived from a system-level analysis of the chip. Other considerations, such as testability, have to be included in the automatic synthesis process. For example, in LSS, there is a redundancy elimination step necessary to meet the testability constraints posed on the design by the rigid rules enforced on IBM computing systems [43].

In the MIS systems [44], timing constraints are also passed to the module generator tools to guide the placement and routing of the gates within a macrocell, and similar constraints are passed to the floor planning and placement and routing tools to guide the placement and routing of the macrocells.

For multi-level design, there are two basic approaches to the logic optimization step:

a) "Global" optimization, where the logic function is re-factored into an optimal multi-level form without consid-ering the form of the original description (e.g., the York-town Silicon Compiler [27], the MIS system [44], part of Angel [52] and SOCRATES [47], and FDS [49]).

b) "Peephole" optimization, where local transforma-tions are applied to the user-specified (or globally opti-mized) logic function (e.g., a part of Angel, LSS [42], MAMBO [51], SOCRATES [47]).

Some global optimization algorithms were proposed in the past (e.g., [45]) to factorize a Boolean function, but these techniques required an exhaustive search which is pro-hibitively expensive for the complexity of control logic that designers are interested in today. Some other algorithms suffered from lack of understanding of the technological constraints associated with a particular implementation of the logic. New algorithms have appeared which are effec-tive in partitioning complex logic functions [46] and can take into consideration the technological constraints of a par-ticular implementation. In addition, rule-based systems [42], [47] have been effective in the actual design of large sys-tems.

1) The Global Optimization Approach: The goal of this step is to reduce the complexity of the logic equations using global techniques which allow significant restructuring of the network, and which are independent of the particular design style or technology. The most difficult and important step in global minimization is to identify factors common to two or more functions which can be used to reduce the total number of literals in the network. This step in this approach is carried out by a set of powerful algorithms which originated from the work in [46]. In this approach, two basic operations are used: generating common alge-braic factors from the equations and checking whether an existing function is a factor of some other functions.

Based on the notion of a kernel [46] the set of all useful algebraic common divisors of a set of equations consists of common single cubes and all intersections of all kernels of each equation. The search space is then restricted to find-ing common single-cube divisors, common kernels, and common intersections of kernels. In the work done at Berkeley on the MIS system, we can demonstrate that these operations are computationally equivalent and can be for-mulated as the problem of finding a minimum cover of a 0-1 matrix by rectangles. Because of the computational complexity of the rectangle covering problem we feel it is not feasible to find the global minimum. Hence, we are forced to use a subset of all divisors in a heuristic manner.

Algebraic techniques fail to exploit any of the Boolean properties of the logic equations. To improve the results, resubstitution is performed, which is to check if any of the existing functions is a divisor (either Boolean or algebraic) of other functions in the network. The usefulness of this operation is supported by the following fact: a function may have a Boolean divisor that has more literals than the func-tion itself, but this divisor is valuable only if it already exists in the current Boolean network. Boolean resubstitution can provide better decomposition, but in general takes much more time than algebraic resubstitution.

To minimize the number of inverters needed to imple-ment the network, global phase assignment is performed which determines for each function whether to implement the function or its complement. This problem can be for-mulated as a constrained column covering problem. In the MIS system, we have exact and fast heuristic solutions.

After global optimization is performed, technology-specific transformations have to be used to ensure that the resulting logic can be implemented in the target technology in the most compact and efficient way. The basic steps in local optimization are: decomposing large gates into smaller ones, deriving better implementations of gates, and simplifying each gate with its local environment.

Different algorithms can be used to decompose a gate into a set of smaller gates. These steps are common to YSC and MIS. "Good algebraic decomposition" uses a greedy strategy based on selecting the best kernel and pulling out this kernel. "Quick factoring" is done by generating only a single level-0 kernel at each step. "Quick factoring" is computationally less expensive than generating all kernels and their intersections, and has been shown to produce good results.

For the complex-gate CMOS design style adopted in the MIS system, another local optimization factors the logic equation of a single gate in order to produce an optimal pull-down and pull-up network for the gate. Different factoring algorithms have been explored. Each has its own use and run-time cost. "Quick factoring" (mentioned above) is useful to estimate the cost of an implementation because the number of transistors in a quickly factored form gives a good estimate of the number of literals in a complex-gate CMOS implementation (and the cost function is evaluated frequently during the synthesis). Two other factoring algorithms are also implemented which give a good algebraic factoring and a good Boolean factoring. Even though they are relatively expensive operations, they yield much better factorizations, and are used to derive the final implementations of gates.

Simplification is used to eliminate redundancy from the network. If all redundancies were removed from the multilevel implementation, then we could guarantee that all the single stuck-at faults in the network would be testable. However, this step is very expensive and a variety of alternative algorithms have been developed, each with different complexity and redundancy-elimination power.

Simplification consists of minimizing the function of a single equation using the algorithms of ESPRESSO-II [26]. Depending on the stage of the optimization, minimization algorithms of differing cost/performance tradeoffs are used. For example, quick simplification [26] is used at the beginning of the synthesis. Later, towards the end of the synthesis, the full power of the ESPRESSO-II minimization algorithm is used with a 'don't-care' set derived from the environment of the function. More details on the algorithms can be found in [33].

To support timing optimization, a set of routines is provided in the global optimization approach to break large gates into smaller ones, or to combine small gates into larger ones, in order to reduce the delay.

The first problem is to estimate the delay of the network. The goal here is a reasonably accurate, relative measure of the speed of the circuit in terms of the number of gates, number of fanouts, and the size of the gates along a critical path. In MIS, we took the following approach to estimate the delay through each gate. First, for each input, we translate a gate of arbitrary series-parallel transistors into an equivalent n-input NAND gate. A NAND gate is characterized by the number of inputs, transistor widths, and load capacitance. The delay is modeled with a polynomial function of these parameters. The coefficient of the equation is determined by the least square fitting of the curve to a large set of SPICE results for various NAND gates [51].

Given a set of timing constraints, the optimization loop involves computing delays through each gate, identifying all the critical paths, finding a minimum weighted node cutset of the critical paths, and the resynthesis of these nodes. Finding a minimum weighted node cutset is equivalent to finding the maximim flow/minimum cut in a flow network. Because of the computational efficiency of the routines to compute delays and cutset, we are able to iterate over these operations many times until the timing constraints are satisfied, or until it is apparent that they cannot be satisfied.

Because the timing estimates are necessarily inaccurate, we view the result of the optimization not as a precise delay calculation, but rather as a resonable guide for restructuring the architecture of the network to meet the timing constraints. Also, the timing estimates are reasonable specifications for synthesizing and placing the gates. More accurate timing estimates and verifications may be employed later in the design cycle when the details of the gate designs and placements are known more precisely.

In a synthesis system, it is important to have a user-friendly, concise way of expressing the logic to be synthesized. The representation of the design should also provide an input for a simulation tool which could verify early in the design cycle the functionality of the design. In the Yorktown Silicon Compiler, a language embedded in APL was used to capture the functionality of the combinational blocks of the design.

In the MIS system, the design is specified using the hardware description language BDS which is part of Digital Equipment Corporation's DECSIM simulation system. Simple extensions to BDS allow for the explicit specification of 'don't care' conditions in the network.

In the version of the logic synthesis system developed at the University of Colorado, the language CSIM [53] embedded in C is used to carry out the task of capturing the design specification.

While the Yorktown Silicon Compiler and MIS are experimental projects with a large research component, the Functional Design System of AT&T Bell Labs is a synthesis system built to speed up the logic design time in a production environment. The synthesis process starts from a high-level description of the design which mixes behavioral and structural representations of the design. System primitives such as registers, counters, adders, decoders, multiplexers, parity generators, and comparators are available to the user who can specify parameters such as number of bits, load, and clocking strategy. System primitives are used by the designer to describe their design structurally. A random logic primitive is available to synthesize multi-level logic described behaviorally using Boolean equations. These equations are manipulated by algorithms which extract common sub-expressions and perform logic simplification. The target logic consists of basic logic gates such as AND-OR-INVERT, OR-AND-INVERT, NAND, NOR, XOR, XNOR, and INVERT. The target logic has fan-in constraints and the decomposition tool maps the Boolean equations into the logic primitives satisfying the constraints. The final step of the design system consists of mapping all the primitives into existing polycells and of placing and routing the cells automatically.

The system has been used on more than 30 MOS chips yielding an estimated 20- to 50-percent reduction of design time. The average design size is about 5000 gates in addition to PLAs and memory. Worth noting is the attention paid to design for testability. The chips designed with FDS have 98-percent fault coverage for the test patterns generated by the system.

The Angel system of NTT [52] is based on similar concepts. The input language allows the description of the behavior of the system to be designed at a higher level than the FDS system. The input description is mapped into an intermediate description which is then optimized. The optimization is accomplished by local transformations as well as by some global factorization algorithms. The final step of the system targets the intermediate representation to a particular technology: Angel supports TTL, ECL, IIL, and CMOS standard cells. This step is performed with local transformations which satisfy technological constraints such as fan-in and fan-out limitations.

This system has been compared to manual designs. For small chips of less than 500 gates, Angel has produced designs which are more compact than manual designs. For larger chips, the efficiency of the algorithms seems to decrease. In fact, for the largest example reported (about 2000 gates), the gate count was increased by 70 percent. The speed of the tool is remarkable, since it has been reported that 10 000 gate circuits can be synthesized in less than 3 MIPS computer.

2) The Rule-Based Approach: The first example of an efficient synthesis system was the Logic Synthesis System of IBM [42]. This system takes a register transfer level representation of the design such as BDL/CS and maps it into a variety of different gate-array technologies. The basic philosophy followed in the system is to apply a number of local transformations, hierarchically organized, to the given description to produce a highly efficient implementation. The emphasis of the system is on producing an acceptable, rather than optimal, implementation in terms of design constraints such as delay, power consumption and area in a short time. For this reason, the local transformation approach has been favored over a global approach.

The highest level of local transformations maps the register-transfer level description into primitives such as ANDs, ORs, INVERTERS, and large macros, such as decoders and adders. The local transformations are mostly translations of high-level constructs into lower level ones and little intelligence is provided, even though some logic optimization is performed. The next level of local transformations consists of transformations which map the previous description into NANDs or NORs. There are about nine rules applied at this level. These rules are responsible for much of the optimization of the system. The transformations used for NAND gates are summarized in Fig. 3. Such local approaches, as used in LSS, tend to be faster than the global schemes but they are somewhat limited in their search for a better design. The final level consists of transformations which map the previous description into technology-specific elements. It is the task of these transformations to take most advantage of the actual technology used.

Throughout all the levels, delays are considered and transformations are triggered if they provide improvement in performance, area, or both. The sequence in which the rules are fired is, of course, crucial for the quality of the final results. The strategy followed is essentially a greedy one where the criterion used is a combination, often specified by the user, of performance and area.

The system has been used on a variety of designs, and is now a production tool. Several problems that arose from

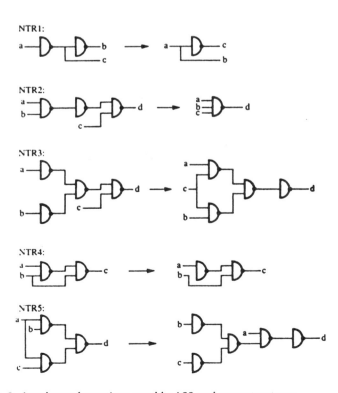

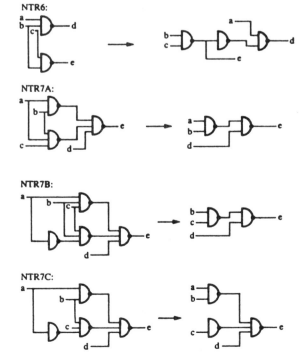

Fig. 3 Local transformations used by LSS and NAND structures.

the application of LSS to production designs had to be solved: testing, path length, incomplete support of the technology, and chip interface information. Of particular interest is the technique used to guarantee testability of the logic generated. This technique identifies the redundancy in the logic [43]. If part of the logic is redundant, then there is no test for that part. Redundancy not only created testability problems, but also large gate counts.

Comparisons with manual designs show that the automatically synthesized designs are within 10 to 20 percent of manual designs in terms of area, but all the constraints were satisfied. One of the most interesting uses of LSS was for technology remapping, i.e., the redesign of the same logic function in a different technology.

In principle, the local optimization approach could be applied to other design styles. In fact, there are at least two projects dealing with local optimization for dynamic and static CMOS cells. MAMBO [51] is a prototype system for the synthesis of multi-level combinational logic in the form of gate-matrix-like domino CMOS logic that has been developed at Berkeley. To characterize this problem, the CMOS domino design style has been characterized using both simulation and analytical techniques to predict the performance of arbitrary, multi-level domino networks. These models of the technology are then used to drive a rule-based logic-level optimization program, MIMIC, which considers not only performance issues, such as circuit speed and area, but also other design constraints including charge redistribution and feedthrough effects.

IV. CONCLUSIONS

The synthesis of efficient digital logic circuits remains a major challenge with the potential for a major payoff. In many ways, new and advanced logic synthesis techniques may provide the "technological advantage" that an understanding of processing technology provided to the successful semiconductor companies of the 1960s.

The most difficult and most time-consuming aspect of the problem today is the synthesis of random, multi-level combinational and sequential control circuitry. Major advances in the area of two-level logic design have been made over the past few years, but we have only just scratched the surface when it comes to multi-level combinational logic design and two-level and multi-level sequential design for integrated circuits and systems.

The important problems highlighted in this paper include;

1) Development of a high-level specification language that supports the capture of 'don't care' information for both sequential and combinational designs.
2) Improved input and output encoding algorithms for two-level logic systems and as a first step in the direction of multi-level logic design.
3) Effective combination of strong and weak division for global, multi-level logic design.
4) The evaluation of combined local and global optimization strategies that take into account the logic function as well as technology-related electrical design constraints and objective functions.

5) New state-encoding schemes designed for an integrated circuit implementation of FSMs.

While a significant amount of research work is ongoing at higher levels of the design process, including behavioral and architectural levels of design, these areas still promise major improvements in design quality and efficiency once effective CAD tools are developed to help system-level designers. This area of research will certainly receive increased attention over the next few years.

The interaction of logic synthesis systems with other parts of a design environment must be well understood. Links with simulators, timing verifiers, logic verifiers, test pattern generation tools, placement and routing, compaction, and the infrastructures, i.e., the database management system and the user interface, have to be provided, and the flow of the design must be well specified to achieve our ambitious goal of designing complex ICs automatically with the same results or better than human designers.

ACKNOWLEDGMENT

Our work in logic synthesis at Berkeley is a group effort. Many people have contributed in essential ways to the understanding of logic synthesis issues and to the development of new algorithms. In particular, we would like to thank Dr. R. Brayton of the IBM T. J. Watson Research Center, whose results have opened new vistas in the field of multi-level logic synthesis. His interaction with the Berkeley group has been invaluable. We would like also to acknowledge the contribution to the Berkeley IC Synthesis Project of Prof. C. Sequin and his students. R. Rudel is the author of many of the logic synthesis tools we have described here. A. Wang and S. Devadas have contributed to the development of new algorithms. P. Moore and D. Harrison have contributed the infrastructural tools on which our synthesis system is based. Our former students Dr. G. De Micheli and Dr. M. Hofmann have contributed in a fundamental way to the development of the early tools.

REFERENCES

[1] Various papers in *Proc. IEEE* (Special Issue on Computer-Aided Design), vol. 69, no. 10, Oct. 1981.
[2] T. Blank, "A survey of hardware accelerators used in computer-aided design" *IEEE Des. Test Comput.* vol. 1, no. 3, pp. 21–43, Aug. 1984.
[3] A. Weinberger, "Large scale integration of MOS complex logic: A layout method," *IEEE J. Solid-State Circuits*, vol. SC-2, no. 4, pp. 182–190, Dec. 1967.
[4] J. W. Jones, "Array logic macros," *IBM J. Res. Devlop.*, pp. 98–109, Mar. 1975.
[5] R. A. Wood, "A high density programmable logic array chip," *IEEE Trans. Comput.*, vol. C-28, no. 9, pp. 602–608, Sept. 1979.
[6] A. Hui *et al.* "A 4.1K gates double metal HCMOS sea of gates array," in *Proc. 1985 IEEE Custom Integrated Circuits Conf.*, pp. 15–21, May 1985.
[7] S. C. Lee, *Digital Circuits and Logic Design.* Englewood Cliffs, NJ: Prentice-Hall, 1976.
[8] C. A. Mead and L. A. Conway, *Introduction to VLSI Systems.* Reading MA: Addison-Wesley, 1980.
[9] D. Johannsen, "Bristle blocks: A silicon compiler," in *Proc. 1st Caltech Conf. on VLSI*, Caltech Computer Science Dep., Pasadena, CA, 1979.
[10] A. C. Parker and S. Hayati, "Automating the VLSI design process using expert systems and silicon compilation," this issue, pp. 777–785.

[11] J. Soukup, "Circuit layout," *Proc. IEEE*, vol. 69, no. 10, pp. 1281–1304, Oct. 1981.

[12] A. R. Newton, "A survey of computer aids for VLSI layout," in *Dig. Tech. Papers*, 1982 Symp. on VLSI Technolgy, (Oiso, Japan, Sept. 1–3, 1982), pp. 72–76.

[13] D. E. Thomas *et al.* "Automatic data path synthesis" *Computer*, Dec. 1983.

[14] T. J. Kowalski *et al.* "The VLSI design automation assistant: From algorithms to silicon" *IEEE Des. Test Comput.*, vol. 2, pp. 33–43, Aug. 1985.

[15] A. R. Newton, "Techniques for logic synthesis," in *VLSI Design of Digital Systems*, *Proc. IFIP TC 10/WG 10.5 Int. Conf. on Very large Scale Integration.* (Tokyo, Japan, 1985), E. Horbst, Ed. Amsterdam, The Netherlands: North Holland, 1986.

[16] A. L. Sangiovanni-Vincentelli, "An overview of synthesis systems," in *Proc. IEEE CICC 85*, pp. 221–225, May 1985.

[17] A. R. Newton and A. Sangiovanni-Vincentelli, "Computer-aided design of VLSI circuits," *IEEE Computer*, vol. 19, pp. 38–60, Apr. 1986.

[18] H. Landman "Automatic layout of optimized PLA structures," M. S. thesis, Dep. Elec. Eng. Comput. Sci., Univ. of Calif., Berkeley, 1981.

[19] L. Glasser and P. Penfield, "An interactive PLA generator as an archetype for a new VLSI design methodology," in *Proc. 1980 Int. Conf. On Circuits and Computers.*, (Rye, NY, Oct. 1980), pp. 608–611.

[20] G. D. Micheli *et al.* "A design system for PLA-based digital circuits," in *Advances in Computer-Aided Engineering Design*, vol. 1, A. L. Sangiovanni-Vincentelli, Ed. Greenwich, CT: JAI Press, 1985.

[21] G. De Micheli, "Symbolic minization of logic functions," in *Proc. Int. Conf. on Computer-Aided Design*, pp. 293–295, Nov. 1985.

[22] S. Devadas and A. R. Newton, "GENIE: A generalized array optimizer for VLSI synthesis," in *Proc. 23rd Design Automation Conf.*, June 1986.

[23] S. J. Hong, R. G. Cain, and D. L. Ostapko, "MINI: A heuristic approach for logic minimization," *IBM J. Res. Develop.*, vol. 18, pp. 443–458, Sept. 1974.

[24] S. Kang, "Automated synthesis of PLA based systems," Ph.D. Dissertation, Stanford Univ., 1981.

[25] D. W. Brown, "A state-machine synthesizer—SMS," in *Proc. ACM/IEEE Design Automation Conf.* (Nashville, TN, Jan. 1981), pp. 301–304.

[26] R. Brayton, G. Hachtel, C. McMullen, and A. Sangiovanni-Vincentelli, *Logic Minimization Algorithms for VLSI Synthesis.* Hingham, MA: Kluwer Academic Publishers, 1984.

[27] R. K. Brayton *et al.* "Automated Implementation of switching functions as dynamic CMOS circuits," in *Proc. 1984 IEEE Custom Intergrated Circuits Conf.*, (Rochester, NY, May 1984).

[28] R. K. Brayton and C. McMullen "Synthesis and optimization of multistage logic," in *Proc. 1984 Int. Conf. on Computer Design*, (Rye, NY, Oct. 1984), pp. 23–30.

[29] R. Rudell, "ESPRESSO II-C user's manual," Dep. of EECS, Univ. of Calif., Berkeley.

[30] ——, "Multiple-valued logic minimization for PLA synthesis," Electronics Res. Lab. Memor. UCB/ERL M86/65, Univ. of Calif., Berkeley, June 1986.

[31] R. Rudell and A. L. Sangiovanni-Vincentelli "ESPRESSO-MV: Algorithms for multiple-valued logic minimization," in *Proc. 1985 Custom Integrated Circuits Conf.* (Portland, OR, May 1985).

[32] A. Poretta, M. Santomauro, and F. Somenzi, "TAU: A fast heuristic logic minimizer," in *Proc. 1984 Int. Conf. on CAD*, (Santa Clara, CA, Nov. 1984), pp. 206–208.

[33] M. Bartholomeus, L. Reynders, and H. De Man, "PLASCO: A silicon compiler for PLA-based systems," in *Proc. 1985 Custom Integrated Circuits Conf.*, (Portland, OR, May 1985).

[34] D. Patterson, "Proceedings of CS290R, Smalltalk on a RISC—Architectural investigations," Computer Sci. Div. Univ. of Calif., Berkeley, Apr. 1983.

[35] C. Marino, "Smalltalk on a RISC—CMOS implementation," M. S. Thesis, Univ. of Calif., Berkeley, May 1985.

[36] T. Sasao, "Input variable assignment and output phase optimization of PLAs," *IEEE Trans. Comput.*, vol. C-33, pp. 879–894, Oct. 1984.

[37] T. A. Dolotta and E. G. McCluskey, "The coding of internal states of sequential machines," *IEEE Trans. Electron. Comput.*, vol. EC-13, pp. 549–562, Oct. 1964.

[38] J. Hartmanis, "On the state assignment problem for sequential machines 1," *IRE Trans. Electron. Comput.*, vol. EC-10, pp. 157–165, June 1961.

[39] G. De Micheli, R. Brayton, and A. Sangiovanni-Vincentelli, "KISS: A program for the optimal state assignment of finite-state machines," in *Proc. 1984 Int. Conf. on CAD* (Santa Clara, CA, Nov. 1984), pp. 209–212.

[40] G. De Micheli, "Computer-based synthesis of PLA-based systems," UCB/ERL M84/31, Electronics Res. Lab., Univer. of Calif., Berkeley, Apr. 1984.

[41] J. Hartmanis and R. E. Stearns, *Algebraic Structure Theory of Sequential Machines.* Englewood Cliffs, NJ: Prentice-Hall, 1966.

[42] J. A. Barringer *et al.* "LSS: A system for production logic synthesis," *IBM J. Res. Devel.*, vol. 28, no. 5, pp. 537–545, Sept. 1984.

[43] D. Brand, "Redundancy and don't cares in logic synthesis," *IEEE Trans. Comput.*, vol. C-32, no. 10, pp. 947–952, Oct. 1983.

[44] R. Brayton *et al.* "Multiple level logic optimization system," in *Dig. Tech. Papers, IEEE Int. Conf. On CAD*, (Santa Clara, CA, Nov. 1986), pp. 356–361.

[45] R. L. Ashenhurst, "The decomposition of switching functions," in *Proc. Int. Symp. on the Theory of Switching*, Apr. 1957.

[46] R. K. Brayton and C. T. McMullen, "The decomposition and factorization of Boolean expressions," in *Proc. 1982 ISCAS Symp.*, (Rome, Italy, May 1982), pp. 49–54.

[47] A. de Geus and W. Cohen, "A rule-based system for optimizing combinational logic," *IEEE Des. Test Comput.*, vol. 2, pp. 22–32, Aug. 1985.

[48] G. De Micheli and A. Sangiovanni-Vincentelli, "Multiple constrained folding of programmable logic arrays: Theory and applications," *IEEE Trans. Computer-Aided Design*, vol. CAD-2, no. 3, pp. 151–167, July 1983.

[49] J. Dussault, C-C Liaw, and M. Tong, "A high level synthesis tool for MOS chip design, in *"Proc. 22nd Design Automation Conf.*, (Albuquerque, NM, June 1984).

[50] C. K. Erdelyi, W. R. Griffin, and R. D. Kilmoyer, "Cascode voltage switch design," *VLSI Des.*, pp. 78–86, Oct. 1984.

[51] M. Hofmann and R. Newton, "A synthesis system for CMOS domino logic," in *Proc. 1985 Int. Symp. on Circuits and Systems* (Kyoto, Japan, June 1985).

[52] T. Hoshino, M. Endo, and O. Karatsu, "An automatic logic synthesizer for integrated VLSI design system," in *Proc. 1984 Int. Customized Ciruits Conf.* (Rochester, NY, May 1984), pp. 356–360.

[53] M. R. Lightner *et al.*, "CSIM: The evolution of a behavior level simulator from a functional simulator: Implementation issues and performance measurements," in *Proc. Int. Conf. on CAD*, pp. 350–352, Nov. 1985.

John A. Darringer
William H. Joyner, Jr.
C. Leonard Berman
Louise Trevillyan

Logic Synthesis Through Local Transformations

A logic designer today faces a growing number of design requirements and technology restrictions, brought about by increases in circuit density and processor complexity. At the same time, the cost of engineering changes has made the correctness of chip implementations more important, and minimization of circuit count less so. These factors underscore the need for increased automation of logic design. This paper describes an experimental system for synthesizing synchronous combinational logic. It allows a designer to start with a naive implementation produced automatically from a functional specification, evaluate it with respect to these many factors, and incrementally improve this implementation by applying local transformations until it is acceptable for manufacture. The use of simple local transformations in this system ensures correct implementations, isolates technology-specific data, and will allow the total process to be applied to larger, VLSI designs. The system has been used to synthesize masterslice chip implementations from functional specifications, and to remap implemented masterslice chips from one technology to another while preserving their functional behavior.

Introduction

The goal of generating an acceptable, technology-specific hardware implementation from a functional specification is not a new one, and it has received much attention in the past. The nature of this problem depends on the level of the functional description, the set of implementation primitives, and the criteria of acceptability. Early work centered on developing algorithms for translating a boolean function into a minimum two-level network of boolean primitives. Extensions were developed for handling limited circuit fan-in and alternative cost functions [1, 2]. But because these algorithms search for minimal implementations they require time exponential in the number of circuits and thus cannot be used on most actual designs.

Other efforts have attempted to raise the level of specification. The DDL work at Wisconsin [2–4], APDL at Carnegie-Mellon University [5], and ALERT at IBM [6] all began with behavioral specifications and produced technology-independent implementations at the level of boolean equations. The results were usually more expen-

sive than manual implementations and did not take advantage of the target technology. For example, the ALERT system was validated on an existing design, the IBM 1800, and the implementation produced required 160% more gates than the manual design [7].

Attempts have been made to produce more efficient logic and to give the designer more control over the implementation [8–10]. This control has resulted in specification language constraints, so that the specification is at a fairly low level and in closer correspondence with the implementation. This necessarily decreases the advantage of an automated approach, bringing it closer to a system for logic entry than for logic synthesis.

Several tools have been developed at Carnegie-Mellon University to support the early part of the design cycle [11–14]. In one experiment [15] the CMU-DA (Carnegie-Mellon University–Design Automation) system was used to implement the data path portion of a Digital Equipment Corporation (DEC) PDP-8/E. It began with a functional

Reprinted with permission from *IBM Journal of Research and Development*, vol. 25, no. 4, pp. 272–280, July 1981.

description of the machine and produced an implementation in two technologies of the registers, register operators, and their interconnections, but not the control logic to sequence the register transfers. When the target technology was TTL series modules the implementation required 30% more modules than the DEC implementation. With CMOS standard cells it required 150% more area than an existing Intersil chip.

There has also been recent work in logic remapping, transforming existing implementations from one technology to another. A group in Japan has described a system to help a designer translate an existing small- or medium-scale integration implementation into large-scale integration [16].

Our approach focuses on the control portion of synchronous machines, since that design is more error-prone than data path design. Thus we assume that all memory elements of the final implementation are identified in the specification; the goal is to generate the combinational logic that computes, on each clock cycle, new values of outputs and memory elements from inputs and the old values of the memories. Also we are focusing on producing random logic implementations, initially for masterslice chip implementations, instead of generating microcode for a control processor or using a programmable logic array. Our initial experiments have been with logic for single chips, so that chip interface information (inputs, outputs, polarities, sender/receiver requirements) was assumed to be specified. The implementations produced by our system are composed of primitives selected from a specified set, connected to satisfy given performance requirements and technology restrictions, and ready to be placed on a masterslice chip.

In a previous paper [17] we described our approach to this form of synthesis; the present paper is an expansion on work reported in [18]. We are not proposing a completely automatic replacement for the manual design process. Instead, we envision an interactive system in which the user operates on a logic design at three levels of abstraction. He begins with an initial implementation generated in a straightforward manner from the specification. He can simplify the implementation at this level, and, when satisfied, can move to the next level. He does this by applying transformations, either locally or globally, to achieve the simplification or refinement. By being able to operate on the implementation at several levels, the user can often make a small change at one level that will cause a larger simplification at a lower level. By limiting the user to directing function-preserving transformations, we can ensure that in all cases the implementation produced will be functionally equivalent to the specified behavior.

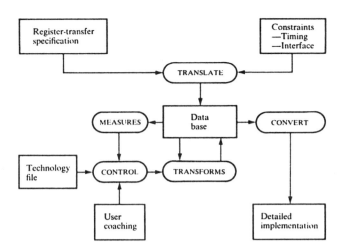

Figure 1 The logic synthesis system.

The use of transformations and levels of abstraction allows a modified form of this scenario to be used in remapping designs from one technology to another. "Remapping" usually refers to the one-to-one substitution of new technology primitives for old technology primitives. Our approach is different: We first transform technology-specific primitives to ones at a higher technology-independent level. To this intermediate-level representation we can apply the synthesis transformations to produce an implementation in a different target technology with the benefit of simplification at several levels.

Both logic synthesis and remapping are problems of finding feasible (not optimal) implementations: networks of primitive boxes that satisfy a large number of constraints. In addition to gate and I/O pin limitations, there are timing constraints, a restricted library of primitives, driver requirements, clock distribution rules, fan-in and fan-out constraints, and rules for testability. Since we hope to apply our techniques to VLSI chips, we are attempting to limit our transformations to local changes that do not require time or space exponential in the number of circuits.

An experimental system for logic synthesis and re-mapping

The organization of the logic synthesis system is shown in Fig. 1. Its inputs are the register transfer specification, the interface constraints, and a technology file which characterizes the target technology. The output is a detailed implementation in terms of the primitives of the target technology, which is submitted to placement and wiring programs for physical design. Some timing or other physical problems may not be detectable before placement and wiring. In this case the synthesis process

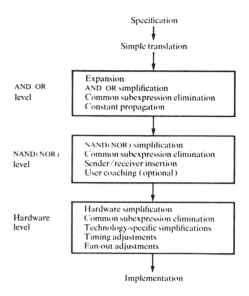

Specification
↓
Simple translation
↓

AND OR level — Expansion / AND OR simplification / Common subexpression elimination / Constant propagation

NAND(NOR) level — NAND(NOR) simplification / Common subexpression elimination / Sender/receiver insertion / User coaching (optional)

Hardware level — Hardware simplification / Common subexpression elimination / Technology-specific simplifications / Timing adjustments / Fan-out adjustments

↓
Implementation

Figure 2 The scenario of synthesis.

is repeated with a revised specification or modified constraints until an acceptable implementation is achieved.

An important requirement of our approach is that the data base be capable of representing the implementation at different levels of abstraction. Our system to support logic synthesis makes use of a graph-like internal data structure for storing the implementation as it progresses from the higher-level description to its final form, and all transformations operate on this graph. There is a single organizational component: the "box." A box has input and output terminals which are connected by wires to other boxes. Each box also has a type, which may be a primitive or may reference a definition in terms of other boxes. Thus a hierarchy of boxes can be used, and an instance of a high-level box such as a parity box can be treated as a single box or expanded into its next-level implementation when that is desirable.

The logic synthesis data base is implemented using a system originally developed for use in an experimental compiler project within IBM Research [19]. It is made up of two groups of tables. The first group describes the technology being used; it is created from a technology file containing for each box type information such as name, function, and number and names of input and output pins. These data are created in batch mode and read during initialization of the interactive system.

The second group of tables contains the representation of the logic created by the interactive system. This group consists of a box table, a signal table, and a set of auxiliary tables which describe the relationship between the boxes and the signals. There is some intentional redundancy in the data: each box has a complete list of input and output signals, and each signal has a source and a list of sinks. Every box table entry contains type information which provides a link to the technology group. This allows programs to get technology information about a specific box.

Transformations communicate with the data base through a layer of functions which perform all data addition, retrieval, and deletion. These functions provide the transformations with the ability to traverse a chip by following signal paths, or by visiting each box. They make it easy to remove boxes and reconnect their input/output signals, to move connections from one box to another, to insert boxes on signal paths, etc. The functions provide a conceptual view of the data base which remains stable even when the data base implementation is altered. The table structure representing this view can be significantly changed with a minimal impact on the processing programs.

The use of data abstraction, of a data base system which allows one to easily define a data base, and of modular implementation of data structures made it possible for us to quickly bring up a usable support system for the transformations. As we learned more about the requirements of the transformations, we were able to change the data base completely, to add and remove data fields, to change individual data structures, and to concentrate efforts in performance improvement in areas where experience indicated that better performance was required. In all cases, only modifications to the data management programs were required to accomplish these changes; the programs which use the data manager were completely unaffected.

The interactive design of the logic synthesis system not only allows the user to control the transform application, but also permits him to invoke programs that aid in his decisions. A BACKTRACE facility displays the cone of influence of a signal or box, showing graphically the logic producing a signal from registers or chip inputs. MEASURE lists, for a design, the number of boxes, signals, connections, inputs, outputs, cells, number of boxes of each type. SEGMENT lists, for each chip output and register, the number and names of the chip inputs and other registers influencing it, the depth of the tree with those leaves, and the number of boxes in it. PRINTBOX lists all boxes of a design and their inputs and outputs, and PRINTREF lists all signals of a design with their sources and sinks. Individual boxes and signals can also be listed in this way. Facilities also exist for producing logic diagrams from a design in the data base.

Expansion and compression commands allow the user to expand a box by replacing it with its more primitive components from a box type definition, and to identify a group of boxes and form a new type of them, replacing the group with a single box. Expansion permits hierarchical development, and compression can be used to partition a design into smaller parts.

The system will accept input in two languages. All of the examples were described in a flowchart-like language, similar to that in [20], allowing GOTOs, assignments to registers and signals, decisions based on the values of registers, computed GOTOs based on values of a group of signals, etc. Parallelism is described in this language by multiple GOTO statements which branch to several actions at the same time. We are also experimenting with a language, similar to CDL (Computer Design Language) [21], that more closely models the internal form of the data base. In addition, it allows convenient description of hardware hierarchy. This aids in the input of box type descriptions which are later to be expanded a hierarchical way, such as a parity function or a decoder.

The synthesis scenario

Though there has been some variation in the synthesis process as the system has been developed and has been applied to more examples, a fairly standard sequence of steps has emerged. Figure 2 shows the three levels of description common to our experiments: the initial AND/OR/NOT level, a NAND or NOR level (depending on the target technology), and a hardware level in which the types of the boxes are books or primitives of the target technology. At every level the implementation is a network of boxes connected by signals. Our objective in devising this scenario was to find a set of transformations and a sequence for applying them such that the original functional specification could be transformed by a sequence of small steps into an acceptable implementation. The transformations at the AND/OR level are local, textbook simplifications of boolean expressions; most of them reduce the number of boxes, but they do not produce a normal form. The NAND and NOR transforms are similar, and required more work because there was less of a foundation on which to build. The hardware transformations were developed after considerable time was spent with chip designers to understand the technologies and the motivation for the many design decisions. Transformations are used not only to simplify the implementation at each level according to appropriate measures but also to move the implementation from one level to the next. The transformations are local in that they replace a small subgraph of the network (usually five or fewer boxes) with another subgraph which is functionally equivalent but simpler according to some measure.

The initial implementation at the AND/OR level is produced by merely replacing specification language constructs with their equivalent AND/OR implementations. Methods for this translation have been described in [3, 5]. At this first level the boxes are of types such as AND, OR, NOT, PARITY, EQ, XOR, DECODE, or REGISTER. Simple local transformations are applied to reduce the number of boxes. Some of the particular transformations used are listed as follows:

$$\text{NOT(NOT}(a)) \Rightarrow a$$

$$\text{AND}(a, \text{NOT}(a)) \Rightarrow 0$$

$$\text{OR}(a, \text{NOT}(a)) \Rightarrow 1$$

$$\text{OR}(a, \text{AND(NOT}(a), b)) \Rightarrow \text{OR}(a, b)$$

$$\text{XOR(PARITY}(a_1, \cdots, a_n), b) \Rightarrow \text{PARITY}(a_1, \cdots, a_n, b)$$

$$\text{AND}(a, 1) \Rightarrow a$$

$$\text{OR}(a, 1) \Rightarrow 1$$

The last two simplifications are examples of a more general constant propagation that is performed. These transformations may leave fragments of logic disconnected. We clean up this disconnected logic in a manner similar to the way compilers perform dead-code elimination. Another technique from optimizing compilers, common subexpression elimination, is also applied here and at other points in the synthesis process to further reduce the size of the implementation. The expansion of "high-level" boxes such as parity and decoders was done here in some of the experiments and was postponed to the hardware level in others. The interactive nature of the system allows this flexibility, which is useful if technology rules require that certain constructs be used for these functions. However, in most cases our simplification rules at the AND/OR and the NOR or NAND levels were powerful enough so that textbook expansions of DECODE, XOR, etc. in terms of AND/OR gates reduced to efficient technology-specific logic.

Next the AND, OR, NOT, and most other operators of the initial description are replaced by their NAND or NOR implementations. The target technologies in our experiments were either NAND- or NOR-based, and this determined the primitive selected for this level. The NANDs or NORs are "idealized," however, in that they have no fan-in or fan-out restrictions. The transition to these primitives is accomplished naively by local transformations, and may introduce unnecessary double NANDs or NORs, which will be eliminated later. Also at this point, the chip interface information is used to place generic (i.e., not technology-specific) senders and receivers on the chip inputs and primary outputs, and to insert inverters where necessary to ensure the correct signal polarities.

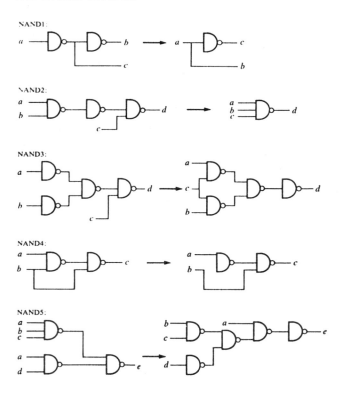

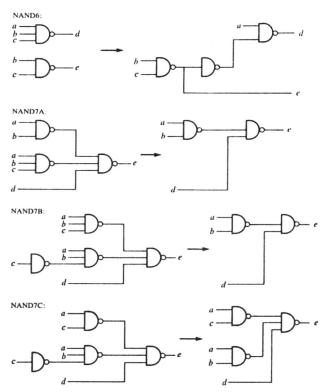

Figure 3 The NAND transformations.

Simplifying transformations are now applied to each signal in the network at this level. These transformations attempt to reduce the number of boxes of the implementation without increasing the number of connections. To accomplish this, the transformations must check the fan-out of the various signals involved, since this will affect the number of boxes and signals actually removed. The transformations are applied repeatedly throughout the network until no more apply. Figure 3 illustrates the NAND transformations used in our experiments; the NOR transformations are identical except for the operator. Each transformation has an associated condition that determines if the replacement will simplify the implementation by reducing boxes or connections. These conditions depend on the fan-out of the intermediate signals and on whether the target technology is assumed to have dual-rail output. For example, NAND3 is only profitable in certain cases. It does not appear to reduce the box or connection count, but if dual-rail outputs are assumed, the single-input NAND on the right-hand side is "free" and disappears after hardware generation. NAND5 is the dual of NAND3, but the two do not cycle because of restrictions on their application. Though NAND5 and NAND6 appear to increase box count, they decrease connections and leave box count the same if dual-rail is assumed.

In the transition to the hardware level, the NAND or NOR gates and generic registers are replaced by technology-specific primitives. Single primitives or macros are selected to match the fan-in of the actual primitives with that of the "idealized" boxes. Also the number of control and data lines of the idealized registers might exceed those normally available, necessitating the generation of additional logic. At this point the implementation is in terms of primitives used by the engineers in their implementations, but because transformations have been made locally there may be some violations of timing, fan-out, and other technology restrictions.

The simplifying transformations at the hardware level are of two sorts. Some are simplifications similar to those at the previous levels, such as eliminating the equivalent of double NOTs, which may occur as a result of expanding higher-level boxes. Others attempt to take advantage of the particular technology. For example, flip-flops may provide an output and its complement, allowing some inverters to be removed at this level. Also, because of combination flip-flop-receiver books available, some receivers may be eliminated. Wired or dotted ANDs or ORs can be introduced to reduce cell count where possible. Some technologies may be dual-rail, having both phases available at every gate; this makes possible simplifications not possible with the technology-independent earlier levels. Other technology-specific transformations applied at this level distribute clock signals to flip-flops according to the technology rules, eliminate long and short paths between flip-flops (assuming a unit gate delay and technology-specific guidelines), and adjust fan-out by repowering signals.

Several of the transforms at the three levels are analogous, differing only in the types of boxes to which they apply, so that simplifications not made at one level would be caught later. This may appear redundant; however, the application of transforms as early as possible reduces the size of the implementation and helps prevent a greater explosion in size when, for example, conversion to NANDs takes place. Though the same implementation might be produced without the NAND simplifications, they are included for efficiency.

The expansion of boxes in terms of more primitive gates was first done only at the hardware level. However, in successive experiments it was found that expansions at other levels were sometimes desirable. For example, if a counter could be expanded in terms of ANDs and ORs, the same expansion could be used for various technologies. The expansion transform therefore was extended to permit selective expansion of box types at various levels.

Synthesis experiments

The synthesis system has been used to create several chip implementations in two different technologies. In some cases, an engineer had implemented the same chip, and we were able to compare the automated design with that of the engineer. In other cases no implementation had been previously attempted.

The first experiments with the logic synthesis system were attempts to produce implementations for chips from existing processors that had been specified functionally and implemented by engineers. The existence of the engineers' implementations permitted comparison of designs and a study of the differences between manual designs and those produced automatically. Each of the experiments was carried out automatically, although the particular sequence of transformations was the result of much experimentation.

• *Experiment 1*
For our first experiment we selected a straightforward chip that had already been manually designed. The specification described seven registers totaling 24 bits, two parity operators, and the conditions for the data transfers. The target technology was a TTL masterslice that provid-

ed 96 I/O pins and 704 cells (divided between three- and four-input NAND gates) on each chip. In addition to the NAND gates, there are a number of macros such as receivers, senders, and flip-flops that are implemented with these NAND gates. Restrictions on the use of the primitives available, such as fan-in and fan-out requirements, timing constraints, clocking and powering rules, were described in the technology file or in some cases built into the transformations. In this experiment EQ, XOR, PARITY, and other high-level boxes were not expanded until the hardware level.

In examining the implementation after the NAND transformations were applied, it was noticed that further improvements could be made. In particular, a reduction in fan-out of a signal by repowering its source would allow a transformation to apply and eventually reduce the size of the implementation. The system allows repowering and some other transformations to be applied to particular signals, rather than across the whole implementation, as a form of user "coaching." In this instance coaching saved only four boxes, but resulted in an implementation slightly better than the manual design.

The first experiment resulted in a synthesized implementation that was remarkably similar to the manual one. In fact, it required four fewer cells, five fewer connections, and four shorter paths than the engineer's implementation. The similarity, however, was not such a surprise since we had used this example in the design of our system, and since we had worked so closely with the chip's designer.

• *Experiment 2*
In the second experiment the same sequence of transformations was applied to a more complex chip. The chip specification contained 13 register bits, a three-bit counter, a five-bit counter, two parity operators, and more complex conditions controlling the data transfers. The target technology was the same as in the first experiment. This time there was virtually no contact with the engineer who designed the chip.

While we tried to use the same scenario, we did make two changes. There was no coaching in this experiment and counters were handled differently from the EQ and PARITY in the first experiment. We found that it is better to expand the counters at the AND/OR level than at the hardware level. This exposes the expanded counter to all subsequent simplifications and allows one definition to be used for different technologies. The expansion transformation therefore has been extended to permit expansion of a nonprimitive box at any level.

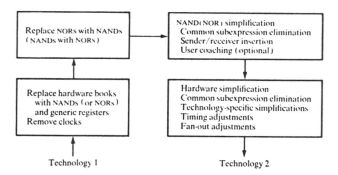

Figure 4 The scenario for remapping.

The synthesis of the second chip resulted in an implementation with 15% more cells and 20% more connections than the manual implementation. We are currently analyzing these results to understand why our implementation is more complex.

• *Experiment 3*

The third experiment was an attempt to synthesize another complex chip in a different technology. This third chip specification described 28 register bits, three parity operators, four decoders, seven comparators, and even more complex control logic. The target technology was an ECL masterslice. In addition to a new set of technology rules and restrictions, this meant that the basic primitive was a NOR and that each primitive had "dual-rail outputs"; that is, it provided both polarities of its output. The synthesis scenario was adapted to this technology and changed slightly, but the three levels of implementation were maintained. The decoders and comparators were expanded at the AND/OR level and the AND/OR transformations remained unchanged. Common subexpression elimination was applied more often at this level and throughout the scenario.

The NAND level became the NOR level because of the new technology. This required a new transformation to translate the AND/OR primitives into NORs, and a set of NOR simplification transformations. These were originally just the NAND transformations with the NANDs converted to NORs, but we later realized that with dual-rail outputs, an apparent box saving at the NOR level might not be a saving at the hardware level, and that the transformation might increase fan-in or number of connections. Thus different fan-out restrictions were used in the NOR transforms. The technology-specific transformations had to be rewritten for the new technology, and some new ones were added, such as the one to eliminate inverters.

This experiment resulted in an implementation with 5% more gates than the manual one. We are trying to account

for this additional logic and determine if it could be eliminated through local transformations.

The remapping scenario

The logic synthesis system has been used to remap chips from one technology to another. Our approach to remapping is not to attempt a one-to-one mapping of hardware primitives, but first to abstract from the hardware level to the technology-independent NAND or NOR level, with generic registers, drivers, and receivers. The NANDs (or NORs) can be mapped to NORs (or NANDs) in a straightforward way, and the NAND/NOR and hardware parts of the synthesis scenario can be applied to produce an implementation in the target technology. This required two new transformations, one that transformed primitives at the hardware level back to the NAND level, and a second that transformed the NAND implementation into a NOR one, while preserving the chip input/output behavior. This approach is better than the straightforward replacement of old technology primitives by new ones, since it exposes the remapped implementation to the simplifications at the NOR level and at the hardware level. Figure 4 outlines the remapping scenario.

Remapping experiments

The first experiment performed was to transform a chip implementation from a TTL masterslice to an ECL masterslice. The chips were of comparable capacity and this chip-to-chip remapping was possible. Since this chip conversion had not been performed manually we could not make an objective comparison. We did check that the input/output behavior was preserved and showed the implementation to an experienced engineer, who found no serious problems.

Chip-to-chip remapping is rare. Usually a new technology will have a different density and number of pins. This could require a merging of several chips from the initial implementation and a partitioning of that remapped, larger function into the chips of the target technology.

Observations

• *Comparing implementations*

One of the problems that confronts us is the difficulty of evaluating the result of the synthesis process. In our work to date, this evaluation has meant a comparison between our generated implementation and a manually produced implementation. There are two aspects to the comparisons that we must perform. One is the problem of determining functional equivalence between the two implementations. The other is to furnish a response to the ill-posed question: "How do these implementations differ?"

Functional equivalence in its full generality is the problem of boolean equivalence and is known to be co-NP complete. This implies that at our present level of understanding it is not possible to devise a program which will efficiently, in all cases, decide equivalence between two implementations. In our case, the problem is often complicated by "don't care" conditions—certain combinations of inputs may be known not to occur. We cannot solve the functional equivalence problem, but we are exploring heuristics which may offer significant speed-up on a large class of implementations. A report on this work is in preparation [22].

Even when two implementations are functionally equivalent, we are still interested in their structural similarity. This form of comparison permits us to evaluate a stylistic difference between our implementation and that produced by an engineer. This is necessary for discovering new heuristics. For this form of comparison we are considering formalizing the notion of "distance" between two implementations, following an analogy to the spelling correction problem.

• *Completeness and coaching*

A desirable property of a set of transformations is completeness—it should be possible to reach any NAND realization of a boolean function from any other by application of the transformations. Our set of NAND transformations does not have this property. Any set of transformations complete in this sense must allow application of transformations in the reverse direction, and this would prevent an automatic application of transformations throughout a design from terminating. What seems desirable is a complete set of bidirectional transformations, with a set of preferred (*e.g.*, box-reducing) directions, yielding a set which terminates with a "good" implementation. The reverse directions would also be available, but only in a user "coaching" mode—they could be invoked on particular parts of the design.

The desire to avoid user-invoked transformations leads to the development of more complicated criteria under which a transformation is to be applied. For example, the coaching described in the first experiment invoked a transformation which would, if applied uniformly, increase the number of boxes in the design. Allowing it to be applied at a particular place by the user has the advantage of providing the (eventual) design improvement desired in the particular case while avoiding building into the transformation constraints on its application. Such constraints may sometimes be worthwhile, but they will make the transformation less local by requiring examination of a larger part of the logic.

• *Technology-specific information*

The technology file allows some generic transformations to apply at all levels of the synthesis process by testing the function of a box to which a transform is to apply, rather than its box type (which may be a hardware primitive). For instance, though it may be necessary to apply a double inverter removal at all three levels, the same transform can be used to do this for NOT, NAND, NOR, and various hardware primitives. A more ambitious use of the technology file would be in hardware generation. For example, a four-way NAND with one input receiving an off-chip signal could be translated by looking in the technology file for a primitive in the target technology implementing that function. It appears that some transformations with specific hardware information built in, such as clock distribution tree generation, will always be necessary.

Future work

Our plans include further analysis of the results of our experiments to determine what improvements should be made to our system. We will also look at more ambitious chips—chips that have required minimization or that have caused long path problems when implemented manually. We hope to arrive at a set of measures and transformations that will provide acceptable implementations for a large class of examples. In addition, we will explore the following:

• multi-chip synthesis—starting with a functional specification that requires several chips, developing additional measures and transformations that will trade resources across chip boundaries.
• engineering changes—examining how such a synthesis system could respond to engineering changes where minimum, local changes are highly desirable.
• transformation specification—looking at how transformations could be described at a high level and compiled for efficient application.
• transformation correctness—considering what properties of transformations (such as function-preservation) should be proved and demonstrating how such proofs can be accomplished.

Summary

We are in the process of exploring what we believe is a new approach to the old problem of logic synthesis and are encouraged by our initial experiments. We have built an experimental synthesis system and used it to synthesize several masterslice chips. In the cases in which we were able to compare our results with previous manual implementations, we found that the automatically produced ones required 0% to 15% more logic. The results are similar when comparing numbers of signals or numbers of connections. We have also used our system to remap implemented chips into a new technology, while preserving their input/output behavior. We plan to perform further experiments, to study the remaining differences between the automatic and manual implementations, and to improve the competence of our experimental system. Our hope is that computationally manageable techniques based on local transformations can be applied to improve naive implementations to acceptable ones. This could greatly shorten processor development and validation times.

Acknowledgments

We would like to thank William van Loo and James Zeigler for many helpful discussions on masterslice chip design, and James Gilkinson for the benefit of his experience in remapping. Also, John Gerbi, Thomas Wanuga, and Alan Stern have made valuable contributions to the design and implementation of the experimental synthesis system.

References

1. M. A. Breuer, Ed., *Design Automation of Digital Systems*, Prentice-Hall, Inc., Englewood Cliffs, NJ, 1972.
2. D. L. Dietmeyer, *Logic Design of Digital Systems*, Allyn and Bacon, Boston, 1978.
3. J. R. Duley, "DDL—A Digital Design Language," Ph.D. Thesis, University of Wisconsin, Madison, WI, 1968.
4. J. R. Duley and D. L. Dietmeyer, "Translation of a DDL Digital System Specification to Boolean Equations," *IEEE Trans. Computers* C-18, 305-320 (1969).
5. J. A. Darringer, "The Description, Simulation, and Automatic Implementation of Digital Computer Processors," Ph.D. Thesis, Carnegie-Mellon University, Pittsburgh, PA, 1969.
6. T. D. Friedman and S. C. Yang, "Methods used in an Automatic Logic Design Generator (ALERT)," *IEEE Trans. Computers* C-18, 593-614 (1969).
7. T. D. Friedman and S. C. Yang, "Quality of Designs from an Automatic Logic Generator (ALERT)," *Proceedings of the Seventh Design Automation Conference*, San Francisco, CA, 1970, pp. 71-89.
8. H. Schorr, "Toward the Automatic Analysis and Synthesis of Digital Systems," Ph.D. Thesis, Princeton University, Princeton, NJ, 1962.
9. C. K. Mesztenyi, "Computer Design Language Simulation and Boolean Translation," *Technical Report 68-72*, Computer Science Department, University of Maryland, College Park, MD, 1968.
10. F. J. Hill and G. R. Peterson, *Digital Systems: Hardware Organization and Control*, John Wiley & Sons, Inc., New York, 1973.
11. M. Barbacci, "Automated Exploration of the Design Space for Register Transfer Systems," Ph.D. Thesis, Carnegie-Mellon University, Pittsburgh, PA, 1973.
12. D. E. Thomas, "The Design and Analysis of an Automated Design Style Selector," Ph.D. Thesis, Carnegie-Mellon University, Pittsburgh, PA, 1977.
13. E. A. Snow, "Automation of Module Set Independent Register-Transfer Level Design," Ph.D. Thesis, Carnegie-Mellon University, Pittsburgh, PA, 1978.
14. L. J. Hafer and A. C. Parker, "Register-Transfer Level Digital Design Automation: The Allocation Process," *Proceedings of the Fifteenth Design Automation Conference*, Las Vegas, NV, 1978, pp. 213-219.
15. A. Parker, D. Thomas, D. Siewiorek, M. Barbacci, L. Hafer, G. Leive, and J. Kim, "The CMU Design Automation System—An Example of Automated Data Path Design," *Proceedings of the Sixteenth Design Automation Conference*, San Diego, CA, 1979, pp. 73-80.
16. S. Nakamura, S. Murai, C. Tanaka, M. Terai, H. Fujiwara, and K. Kinoshita, "LORES—Logic Reorganization System," *Proceedings of the Fifteenth Design Automation Conference*, Las Vegas, NV, 1978, pp. 250-260.
17. J. A. Darringer and W. H. Joyner, "A New Approach to Logic Synthesis," *Proceedings of the Seventeenth Design Automation Conference*, Minneapolis, MN, 1980, pp. 543-549.
18. J. A. Darringer, W. H. Joyner, L. Berman, and L. Trevillyan, "Experiments in Logic Synthesis," *Proceedings of the IEEE International Conference on Circuits and Computers ICCC80*, Port Chester, NY, 1980, pp. 234-237A.
19. F. E. Allen, J. L. Carter, J. Fabri, J. Ferrante, W. H. Harrison, P. G. Loewner, and L. H. Trevillyan, "The Experimental Compiling System," *IBM J. Res. Develop.* 24, 695-715 (1980).
20. G. L. Parasch and R. L. Price, "Development and Application of a Designer Oriented Cyclic Simulator," *Proceedings of the Thirteenth Design Automation Conference*, San Francisco, CA, 1976, pp. 48-53.
21. Y. Chu, "An ALGOL-like Computer Design Language," *Commun. ACM* 8, 607-615 (1965).
22. C. L. Berman, "On Logic Comparison," *Proceedings of the Eighteenth Design Automation Conference*, Nashville, TN, 1981 (to appear). Also *Research Report RC5342*, IBM Thomas J. Watson Research Center, Yorktown Heights, NY, 1980.

Received August 22, 1980; revised January 15, 1981

The authors are located at the IBM Thomas J. Watson Research Center, Yorktown Heights, New York 10598.

A Rule-Based System for Optimizing Combinational Logic

Aart J. de Geus and William Cohen
General Electric Microelectronics Center

T wenty to 50 percent of the active area of most semicustom integrated circuits is devoted to combinational logic. Synthesizing this circuitry from functional specifications is a relatively straightforward process. Optimizing either the size or the performance of such circuitry in a given technology, however, is considerably more difficult and can consume valuable design time. For this reason, manual optimization is often not even attempted except on the most critical portions of a design, leading to chips that are unnecessarily large and slow. In addition, when an existing design is converted from one technology to another, designers have to re-optimize the existing implementation to take full advantage of the target technology.

Automating the synthesis and optimization of combinational circuitry can result in significant improvements in both the design cycle time and the overall quality of the implementation. Automatic synthesis also guarantees

functional correctness. Standard techniques[1] for performing logic level reduction are a major step in this direction, but fail to address the actual technology-dependent, circuit-level implementation. Such minimizers will find an optimal implementation using NAND/NOR logic, for example, but cannot easily take advantage of other gates available in gate-array or standard cell libraries.

Our approach to synthesizing and optimizing combinational logic consists of three steps:
- minimizing the Boolean equations,
- synthesizing an initial network, and
- optimizing the network for a given technology.

During the minimization phase, the set of Boolean equations describing the desired functions is reduced using mathematical methods that take maximum advantage of the "don't care" set. In the second phase, the equations are factored to take advantage of common intermediate terms, and an initial implementation is created by using a limited set of gate types, such as NAND/NOR gates and multiplexors. In the final phase, this network is optimized to take greater advantage of the target technology by performing a series of *local transformations* on the circuit. These transformations are formulated as rules to be applied to the circuit by a rule-based expert system. This article

Summary

SOCRATES is a rule-based expert system that optimizes combinational logic for a specific target technology. The system performs substitutions of equivalent gate configurations, thereby reducing the overall area of the implementation and improving the speed of the design. A control mechanism uses various backup strategies to choose the rules applied to the circuit. Users can easily extend the library of transformation rules through a rule generation module that automatically encodes rules and inserts them into the knowledge base. Timing constraints placed on the circuit can be modified to allow the designer to explore a large design space in a matter of minutes. Implementations generated by the system are comparable in area and speed to circuits designed by experts.

Reprinted from *IEEE Design and Test of Computers*, vol. 2, no. 4, pp. 22-32, Aug. 1985.

focuses on the third phase, the program SOCRATES (Synthesis and Optimization of Combinatorics using a Rule-based And Technology-independent Expert System), which optimizes gate-level circuits for speed and area in a given technology.

System description

Combinational logic is generally represented in one of two forms: as a set of Boolean equations or as a netlist of interconnected gates. The set of Boolean equations, which can be in sum of products or multilevel form, is technology-independent and lends itself to mathematical minimization. The netlist represents a gate-level implementation of the logic and is technology-dependent because its components are limited to the set of gates available in the technology in question. The quality of an implementation is measured in terms of the area and speed of the design. This article does not address the optimization of transistor-level implementations; only gate-level implementations, such as those typically found in semicustom design approaches like gate arrays and standard cells, are considered.

Numerous tools are available that minimize and implement combinational logic. Most of these tools apply at the Boolean level, are technology-independent, and generally assume an AND-OR or NAND-NOR implementation. Such tools fail to take full advantage of the various types of gates available in a semicustom library. For example, these tools will construct a simple exclusive-OR function using NANDs and NORs even if an exclusive-OR gate is available.

The need for more flexible and technology-oriented tools was recognized by Darringer, et al.,[2,3] who implemented a design system allowing users to easily perform local transformations at various levels of abstraction. SOCRATES allows the system, rather than the designer, to perform the transformations. The SOCRATES system is illustrated in Figure 1.

Transformation modules. The Boolean format contains a description of the logic in terms of Boolean equations such as $F = ab + b'd'e + ed$. The equations can be multilevel and a *flattener* is available to put the function in sum of products form. The Boolean format is technology-independent.

The netlist format contains a description of the logic as implemented using a specific set of gates and is technology-dependent.

The *minimization* module reduces the set of Boolean equations describing the logic by using heuristics that find a minimal set of prime implicants. In finding this minimal set, the module takes advantage of the "don't care" sets of the function. The Espresso-IIC[1] program performs the reductions.

Espresso-IIC simplifies the two-level representation of the function. Another tool, which uses a technique known as weak division, takes a two-level function and creates a multilevel function based on small subexpressions that occur often in the original functions.[4] This tool thus detects and eliminates multiple occurrences of the same subexpression, which would otherwise result in duplicate logic in the synthesized circuit.

The *synthesis* module translates a Boolean function into a gate-level implementation. Two synthesis modules have been studied, one that generates an AND/OR implementation of the function and one that generates a multiplexor implementation of the function.[5] The AND/OR implementation gives vastly superior results when weak division is used.

The *extraction* process is the inverse of synthesis. This module generates from a netlist its Boolean function description. The extraction module is often used in conjunction with the *comparison* module, which compares two Boolean functions for functional equivalence. If they are functionally

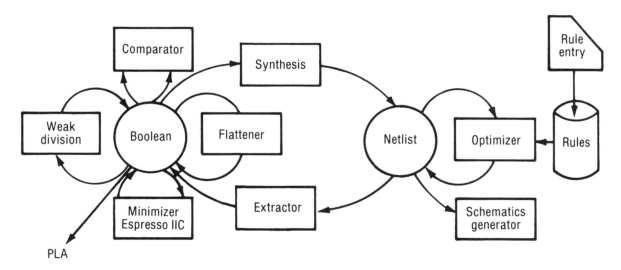

Figure 1. SOCRATES system description.

equivalent, the module outputs a vector of input values at which the two functions differ, allowing the user to isolate the problem. Thus, it is simple to confirm that a succession of various optimization and transformation tools has not altered the functionality of the logic due either to human error or to errors in the code. This verification capability was crucial in developing the SOCRATES system; a number of bugs, both in the knowledge base and in various programs, were quickly detected and corrected.

The *optimization* module performs a succession of substitutions on an existing netlist, similar to the way an experienced designer manipulates a design to achieve greater efficiency. This module consists of a rule-based system that performs substitutions and a rule entry program that helps a user extend the knowledge base.

The SOCRATES system is linked, via the Boolean equation format, to a programmable logic array generation system. Netlists generated by the system can be examined via a schematic generator, which gives the user a crude pictorial representation of a circuit.

Design scenario. Figure 2 illustrates the typical order in which the tools are used. In step (a), the Boolean equations describing an existing netlist are extracted from that netlist; in step (b), the set of equations is minimized; in step (c), weak division is performed to generate a multilevel representation; a new netlist is synthesized in (d) and optimized in (e) for the target technology. If no previous implementation exists, the procedure starts at step (b). Because all the tools interface with each other, either in terms of Boolean functions or in terms of a standard netlist, users can easily develop macros for any desired design sequence. The average user typically follows a standard sequence, whereas more sophisticated users can experiment with the system.

The rule-based expert system

The paradigm of a rule-based expert system has three parts: *data* on which certain manipulations are performed, a *knowledge base* made up of a set of *rules* describing legal manipulations of data and the conditions that have to be satisfied to execute the rules, and a *control module* directing the application of the rules to the data. In addition, the control module of a rule-based system for optimization or design (in contrast to diagnostic systems) requires a cost function to measure the quality of the results obtained. The final component of a rule-based system is some sort of mechanism to extend and maintain the knowledge base. In the SOCRATES system, this role is filled by the rule entry module.

Using knowledge-based expert systems in electrical engineering problems is not new. One of the most successful applications of such systems is the R1[6] system used at DEC to configure large computer systems. Another synthesis system is the DAA (Design Automation Assistant) program[7] developed at Carnegie-Mellon University, which allocates and configures hardware to implement a design from an algorithmic description.

The expert system approach requires that rules remain simple and uniform and that they exhaustively cover the type of data manipulations that an expert would perform. Logic optimization through successive gate-level transformations is thus ideally suited for such an approach.

Knowledge base. SOCRATES optimizes a circuit by performing a series of local transformations to that circuit. In performing each transformation, the program replaces a given configuration of gates by another functionally equivalent configuration of gates. These transformations are always applied in such a way as to reduce the cost function and produce a more optimal circuit. An example of such a transformation rule is shown in Figure 3.

Control module. The control structure of an expert system directs the application of the rules. In SOCRATES the control structure determines what rules or sequences of rules are applicable to the circuit, evaluates which choice is most desirable, and then performs that transformation on the network. Optimization of a combinational network through successive transformations can be translated to the problem of optimally traversing a graph called a *state space*. For SOCRATES, the nodes of this graph are the possible implementations of the circuit, and the arcs represent rule applications. Optimization is thus equivalent to finding a path from the initial circuit configuration to an optimal configuration. The process of finding this path is known as a *state space search*.

Cost function. The implementation's optimality is measured primarily in terms of area. Although reducing circuit complexity usually results in improved circuit timing, an optimization strategy based solely on a measure of the expected chip real estate is not adequate. The approach taken in SOCRATES allows the user to impose timing constraints. The system first

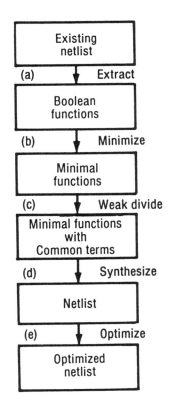

Figure 2. Typical design scenario.

The flow diagram shows:

Existing netlist
(a) ↓ Extract
Boolean functions
(b) ↓ Minimize
Minimal functions
(c) ↓ Weak divide
Minimal functions with Common terms
(d) ↓ Synthesize
Netlist
(e) ↓ Optimize
Optimized netlist

satisfies these constraints, then optimizes for area under these constraints.

Main operations. Four operations are performed during optimization:

• *Matching:* This function finds all or a number of rules that apply to the present network. Implementation of this function is explained in detail in the section discussing the SOCRATES knowledge base.

• *Cost, function evaluation and optimization diagnostics:* This function evaluates the cost function improvement associated with the application of a rule. It also computes a number of parameters affecting the rule selection mechanism, such as the number of rules that can be applied subsequent to the rule in question. Detailed timing information is made available describing the longest path in the circuit and the timing slack of all elements in the network in reference to the longest path.

• *Selection:* This function determines which rule should be applied next on the network. The selection mechanism is further explained in the section on search strategies.

• *Replacement:* This function applies the rule chosen by the selection routine and performs the network transformation. The implementation of this function is explained in detail in the following section.

The knowledge base

Format and structure. In the SOCRATES system, a rule is a mechanism to replace a portion of a circuit by a functionally equivalent but more desirable circuit portion. Rules are stored as a pattern describing a *target configuration* to be recognized in the circuit and an associated action detailing how to build the *replacement configuration*. Substituting the replacement configuration for the target configuration either reduces the overall area of the circuit or improves its performance.

The knowledge base is structured in several ways. First, the set of rules is subdivided into three classes: *general rules,* which dramatically reduce the

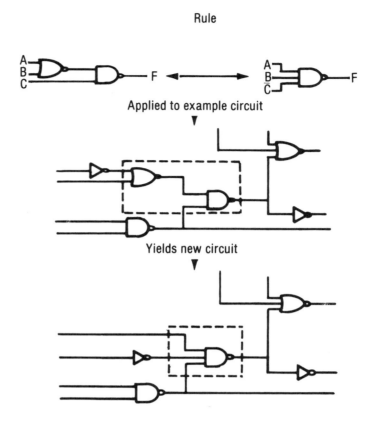

Figure 3. Transformation rule example.

complexity of the circuit by reducing both area and timing; *timing rules,* which improve performance, possibly at the expense of area; and *area rules,* which reduce area, possibly at the expense of timing. The intersection of these classes need not be empty; for example, any transformation that can appropriately be included in both general and area rules will be included in both classes.

The rules within each class are ordered by their relative desirability: timing rules are ordered by the expected savings in performance, area rules by the expected savings in area, and general rules by a value that is a function of both these expected savings. This ordering is necessarily approximate, since a rule's cost function improvement depends on where in the circuit the rule is applied.

Finally, the knowledge base is struc-

tured by using *subclasses.* Optimizations performed by the rule-based system can be divided into discrete and independent phases. For instance, SOCRATES might first eliminate redundant multiplexors, then replace partially redundant multiplexors with two-input NAND and NOR gates, and finally perform a series of transformations that combine and simplify these gates. To improve the efficiency of a search through the knowledge base, the rules that compose each discrete phase of optimization are grouped into subclasses.

Matching. This section explains how the pattern of gates describing a target configuration is detected in the circuit. Each pattern is encoded as a conjunction of *conditions* that must be satisfied. The gates and pins that satisfy these conditions are stored in a set of registers: the *matcher gate registers*

26

and the *matcher pin registers*. Both sets of registers can be accessed by the replacement module. When the pattern matcher is called, it is given two arguments: a pattern to detect, and a gate at which the pattern match should begin. This gate is loaded into the first of the matcher gate registers and provides a reference point from which to start the pattern match.

Figure 4 shows a sample pattern and a portion of a circuit that matches the pattern. In matching the pattern shown, the pattern matcher first checks that the gate stored in gate register G_0 is a two-input NOR gate. If so, then the matcher selects some pin of N_0 and stores it in pin register P_1. If P_1 is an input of G_0 with only one other connection in its net and P_1 is positively connected (i.e., through an even number of inversions) to the output of a two-input NAND gate, then the pattern matcher proceeds through the pattern, saving pointers to the inputs of the NAND gate in pin registers P_3 and P_4. The pattern matcher then returns to gate G_0 and tries to find another input pin that connects to a two-input NAND gate. If at some point a condition cannot be met, the pattern matcher *backtracks* by returning to the last selection it made and making a different choice. For instance, if the first pin it stored in P_1 did not connect to a NAND gate, then line 9 would fail, and the pattern matcher would backtrack to line 2 to find a different value for P_1.

Backtracking continues until the last condition has been satisfied and the pattern is matched, or until all possible choices have been rejected. As this example illustrates, when a pattern is successfully matched, pointers to all the inputs and outputs of the recognized configuration are stored in the pin registers and pointers to the gates that comprise the configuration are stored in the gate registers. These pointers are used by the replacement module when linking in the replacement configuration. In technologies that include such a gate, SOCRATES replaces the pattern shown in Figure 4 with a special AND-OR-INVERT gate.

There are also conditions that check if certain timing conditions hold; for instance, it can be checked if a pin is the slowest input of a gate or if a network's capacitance exceeds a given quantity.

Disjunction, or the OR-ing together of conditions, is not allowed in patterns. A large part of the functionality of disjunction is provided by the use of gate and pin classes. When describing the gates available in a technology, the user can impose a taxonomy on the available gate types and on the pins of each gate. The pattern matcher can then test for membership in any class.

Patterns are also made more versatile by generalizing the handling of inversion. The pattern matcher does not explicitly look for inverters when matching a pattern; this feature significantly increases the scope of each

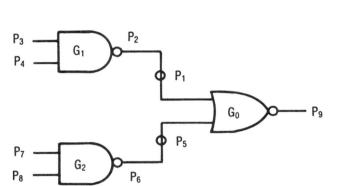

1	node_class	g0	nr2
2	pinof	g0	p1
3	pinclass	p1	nr2_inputs
4	net_size	p1	1
5	con_pos	p1	g1
6	assign_con	p2	g1
7	pin_class	p2	nd2_output
8	net_size	p2	1
9	node_class	g1	nd2
10	pinof	g1	p3
11	pinclass	p3	nd2_inputs
12	pinof	g1	p4
13	pinclass	p4	nd2_inputs
14	pinof	g0	p5
15	pinclass	p5	nd2_inputs
16	net_size	p5	1
17	con_pos	p5	g2
18	assign_con	p6	g2
19	pin_class	p6	nd2_output
20	net_size	p6	1
21	node_class	g2	nd2
22	pinof	g2	p7
23	pinclass	p7	nd2_inputs
24	pinof	g2	p8
25	pinclass	p8	nd2_inputs
26	pinof	g0	p9
27	pinclass	p9	nr2_output

Figure 4. A sample pattern.

rule and thus reduces the number of rules required for satisfactory optimization.

Replacing. After a pattern is matched and pointers to the identified gates and pins are stored in the matcher gate registers and matcher pin registers, the replacement module is called to perform the transformation. This module performs the replacement in three steps. First, a replacement configuration—which can be empty—is built. Second, a series of connections are made that link the replacement configuration to the existing circuit. The next step is deleting the target configuration; this third step also includes a cleanup phase, in which superfluous gates (such as inverters that were looked past by the pattern matcher, and are now unconnected) are removed from the circuit. The final replacement step is performed only when a gate is replaced by two or more parallel gates to increase its drive. In this case, the output load is distributed as evenly as possible between the new drivers.

Search strategies

Problem description. As noted above, optimization can be viewed as a state-space search problem. SOCRATES approximates a gradient search in traversing the search space, always applying the transformation that maximally decreases the cost function.

The first version of the system selected rules based on the ordering of the knowledge base, always applying the first applicable rule in the knowledge base. The basic algorithm was:

```
select_rule:
for each rule R in some class C
   for each gate G in the circuit
      if the target for R matches at gate G
         apply_rule (R, G)
         go to select_rule
```

The value of C controlled what parameters were optimized. For instance, when optimizing for area, the class of *area rules* was used.

This simple control strategy was fairly effective for area optimization and fairly poor for timing optimiza-

tion. The reason was that the effect of a *timing rule* is difficult to predict, so the ordering of the knowledge base was inexact. Another problem was that this strategy failed to provide for intelligent decisions about where to apply a rule; for example, Figure 5 demonstrates a situation where the same rule can be applied in two separate places with dramatically different results. Finally, it appeared that a gradient search could be better approximated by not looking at the effects of individual rules on the circuit, but by examining the effects applying a short sequence of rules. By doing so, we could relax the requirement that each individual rule reduce the cost function, thus allowing a much greater range of transformations.

The look-ahead strategy. For these reasons, a *look-ahead* scheme was implemented. This strategy entails building a portion of the state-space graph, called a *search tree*, explicitly in memory. This tree is then traversed, and the leaves examined. A sequence of rules leading toward the leaf node with the lowest cost function will then be selected by the control module. Finally, a portion of this sequence of rules is applied to the circuit.

Pruning the tree. A circuit with 100 gates and a knowledge base of 50 rules typically evidenced a branching factor of about 50. Therefore, the search tree necessary to look ahead two rule applications would contain more than 2500 nodes; building and exploring a

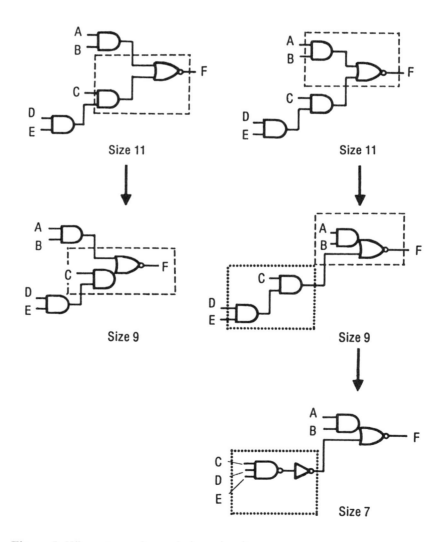

Figure 5. Where to apply a rule in a circuit.

tree of this size with every rule application was obviously impractical.

Three simple heuristics are used to prune this tree. First, the number of sons of any node that are actually explored is limited to some small number B. This number then corresponds to the maximum *breadth* of the tree. The B sons that the control module chooses to explore are the configurations that would be reached by applying the first B applicable rules to the circuit represented by the parent node.

Second, the *depth* of the tree is limited by the parameter D. A depth of three, for example, means that the control module will choose between all reasonable sequences of three rules.

Third, the search tree is restricted to rules that apply to a small area of the circuit. This technique avoids situations where the control module attempts to decide between two rules that can be applied independently of each other. The size of this local area, or *neighborhood,* is determined by the parameter N.

Another parameter that affects the look-ahead strategy is D_{app}, the rule application depth. After the search tree is built and explored, rules are chosen to actually apply to the circuit. These rules are a part of the path from the root of the search tree to the lowest-cost node. The length of this path will be D_{app}; for instance, if $D_{app}=2$, then two rules will be applied to the circuit each time a tree is built.

Metarules. The first look-ahead strategy built and traversed a search tree before applying each rule. The size and shape of this tree was determined by the parameters described above, but was fixed throughout the optimization. A series of experiments fixed the point of diminishing returns for the parameters above at $B=3$, $D=2$, and $N=D_{app}=1$.

Although this simple strategy improved the quality of results by about 10 percent, it was very inefficient. The amount of look-ahead actually needed varies greatly over the course of an optimization; in general, look-ahead is far more useful in later phases of optimization than in earlier phases, and look-ahead is usually superfluous for the most powerful rules. Obtaining the full benefit of look-ahead in the last stages of optimization requires fairly large values of B and D.[8]

Also, the quality of an optimization can be improved by varying the weighting of terms determining the cost function. For example, giving progressively lower weights to the term that reflects the number of rules applicable to the circuit is more accurate than giving it a fixed weight, since the rules applied later in an optimization will not improve the quality of the circuit as much as the earlier rules.

The metarule system acts as a supervisor, ensuring effective use of the look-ahead technique.

For these reasons, some way of dynamically varying these parameters was needed. The *metarule system* is a module that determines appropriate values of the control parameters based on the current state of the control module.[8] The system acts as a supervisor, overseeing the control module and ensuring effective use of the powerful but expensive technique of look-ahead. The metarule system follows the paradigm of a rule-based system. It consists of a series of *metarules,* each of which will adjust a set of control parameters if and only if a given set of conditions holds. The metarule below turns off look-ahead by setting $B=D=1$ for all area-specific rules in subclass 2.

```
*  turn off look-ahead for rules
*      in subclass 2
*

R 15 :=
    ->(rclass == AREA)
    ->(rsubclass == 2)
     ->(look_depth = 1)
     ->(branch_factor = 1)
```

Using the metarule system reduced the runtime of the optimization module by almost 60 percent and modestly improved the quality of the circuits produced, as compared to the older strategy, which used fixed values for the backup parameters.

Building the search tree. Building the search tree requires the following capabilities:

- Selecting an appropriate *locality* for optimization.
- *Expanding* a node in the search tree by finding B sons.
- *Descending* the tree, traveling from a parent to the son.
- *Ascending* the tree, traveling from a node to its parent.
- Traversing the tree, *selecting* a node in the tree, and then putting the circuit in the state represented by that node.

The first step in selecting a locality is choosing a center. This *center* is simply the gate at which the best rule can be applied. The locality will then consist of this node and its neighbors. The parameter N affects how many neighbors are selected. When optimizing for performance, only gates on the critical path are considered as centers.

The search tree, as represented in memory, does not contain an entire copy of the circuit at each node. The tree is only a "road map" describing how to get from one circuit configuration to another via transformations that can be encoded as (rule, gate) pairs. Expanding a node thus consists of finding B (rule, gate) pairs by tracing through the knowledge base and picking out the first B rules that can be applied to the circuit.

Descending the tree is merely applying a transformation to the circuit. Ascending from a son to its parent is more difficult, and must be done by reversing the effects of the last transformation. As transformations are performed, SOCRATES keeps a log of low-level changes to the circuit. Each entry in the log contains enough information to allow reversal of such changes. Periodically, this log is cleared. This approach allows rapid backup; one disadvantage is that the tree must be explored in a depth-first manner.

As each node in the search tree is visited, the cost of implementing that circuit configuration is saved. The final step of selecting the best state is thus straightforward.

Optimization for speed. The above algorithm requires extensions when optimizing for timing. A critical path finder[9] embedded in the optimization system traverses the network, finds the critical path(s) and computes the timing slack at each gate not on the critical path(s); here the timing slack of a gate is defined as the amount of time by which the output signal can be delayed until the signal causes a new critical path. Because this information needs to be recomputed after every application of a rule, the critical path finder allows timing information to be updated incrementally after each transformation, minimizing the computing cost involved.

When optimizing for timing, the control module will only consider rules in the class of *timing rules* and target configurations that include gates on the critical path of the circuit. The system still relies on a gradient search, in which the rule that most improves performance is applied. This search is terminated as soon as the circuit meets the specified timing constraints.

In the last phase, optimization is done exclusively for area. The critical path analysis tool checks that no transformation lengthens the critical path sufficiently to violate the timing constraints; transformations that produce a circuit that is too slow are not considered. While optimizing for area, rules that involve gates with large slack times are given precedence over rules involving gates with small slack times; this feature decreases the number of transformations that must be avoided because of performance considerations.

For most of the optimization process, the control module is restricted to investigating rules that will actually lower the cost of the circuit. Toward the end of the last phase, this requirement is relaxed; at the same time, the

metarule system is used to increase the depth of look-ahead. This gives the rule-based system a final chance to climb out of a local minimum.

Knowledge acquisition

One of the major obstacles to expert systems development has been the problem of knowledge acquisition. In recent years, tools to implement expert systems, such as OPS5[10] and PRO-LOG,[11] have proliferated; however, building the knowledge base of an expert system has remained by and large a black art. Furthermore, testing a knowledge base and verifying its correctness is at best time-consuming; at worst, standards for its correctness

---

One of the major obstacles to expert systems development has been the knowledge acquisition problem.

---

or sufficiency do not even exist.

In the SOCRATES system, the problems associated with knowledge acquisition are potentially severe. First, the knowledge base is highly technology-dependent, and therefore dynamic. Second, the end users of the system are themselves experts, and are not likely to be satisfied with results that do not reflect their own expertise. However, the rules used by the SOC-RATES expert are more constrained in format than those used by many expert systems. Also, a simple standard exists for evaluating the correctness of a transformation: a correct transformation does not change the functionality of the circuit. Finally, the "deep knowledge" necessary to evaluate the relative benefits of applying a rule—the ability to evaluate the performance and area of the two networks—is easily computable.

For these reasons, the SOCRATES expert's domain is ideal for semi-automated knowledge acquisition. The fact that each rule represents a local transformation suggested the

simple input format of two networks, one configuration representing the target configuration and one representing the replacement. The correctness of a rule is easy to check automatically using the extraction and comparison modules. The SOC-RATES module used to semi-automate the knowledge acquisition process is the *rule entry* module.

The rule entry module. Input to the rule entry module is a pair of netlists entered by the user. The first netlist represents the target configuration and and the second netlist represents the replacement configuration. The steps performed automatically by the system during rule entry are listed below:

- *Verifying* the functional equivalence of the two netlists.
- *Generating* a pattern describing the target configuration and an action describing how to replace it with the replacement configuration.
- *Classifying* the new rule into the appropriate class and subclass.
- *Inserting* the new rule into the knowledge base.

The verification step uses existing modules. The functions implemented by the two netlists are extracted and compared; if they are not equal, the rule entry module aborts with an error message.

To generate the pattern describing the target configuration, the netlist is traversed in the same order that the pattern matcher would traverse it and a stream of conditions that must be met is emitted. The conditions are guaranteed to be sufficient to ensure that the pattern matcher recognizes a circuit portion functionally equivalent to the given netlist. Command-line arguments to the rule generation module can be given to control the strictness of the conditions. For instance, the generation phase can produce a pattern that specifies either the gate class of each gate encountered or the actual gate type.

The action associated with the rule is generated by generating commands to build the replacement configuration, linking together pairs of pins that are logically equivalent to the same input or output, and finally deleting the gates found by the pattern matcher in identifying the target configuration. Some optimizations of this action are done: for instance, inversion of inputs and outputs is treated in such a way as to avoid building unnecessary inverters.

The action and pattern generated are saved in a test-level version of the SOCRATES knowledge base. As generation proceeds, a characteristic vector is created that contains information about the transformation; for example, the maximum and minimum potential performance and area savings are computed and stored in this vector. The characteristic vector of every rule is saved in a special table in the knowledge base.

The final step in generating a rule is classifying it. The classification process also reorders the rules by their potential benefits. Classifying a rule into the classes of general rules, timing rules and area rules is done by examining the possible benefits of applying the rule from the standpoint of both performance and area. Classifying a rule into subclasses is done by using a file that describes each subclass by the values of the parameters in the characteristic vector. Some of these parameters describe how rules interact and are a function of the entire knowledge base. Thus, rule classification cannot be done incrementally, as can rule generation; the knowledge base must be evaluated and resorted each time a new rule is added. The time required to evaluate and resort the knowledge base currently used by the system is about three CPU-seconds on a VAX 11/780.

Benefits of knowledge acquisition. The rule generation module decreases the time required to enter and test a new rule from 45-90 man-minutes to about three man-minutes. The process of entering a new rule is also simplified so that it can easily be done by a sophisticated user. Since the rule generation module was incorporated, the SOCRATES system has been transported to two new target technologies, and the knowledge base was rewritten to work with two different synthesis modules.

Results

The SOCRATES system successfully automates a time-consuming and error-prone phase of the design process. Combinational circuits that would have taken several days to synthesize and optimize can now be generated in minutes. The user can influence the optimization to reflect constraints on timing and/or area. The utility of the system is increased by the extraction and comparison modules, which allow automatic verification of the final product.

Two factors contribute to the system's efficiency. First, it is written in C language, rather than a general-purpose expert systems writing tool. Low-level routines are thus problem-specific, which generally leads to increased speed. Second, the combinational explosion involved in using a state-space search has been effectively contained by dynamically maintaining the backup parameters at a point of diminishing returns, thus limiting the search space. All modules of the system are fast enough to allow interactive use for small- and medium-size examples. As Table 1 shows, however, even large and complex functions can be optimized quite quickly. The size of the examples is given in transistor pairs (a two input NAND gate, for example, requires two transistor pairs).

Although the system takes full advantage of the target technology, SOCRATES is also highly technology-independent. Porting SOCRATES to a new technology consists of recreating the knowledge base. Often, a core knowledge base can quickly be assembled from rules in existing knowledge bases; these rules can easily be re-entered and reevaluated for the new technology by using the rule entry system. Currently, the SOCRATES

Table 1.
Run times of optimization module.

Example	Inputs/Outputs	Final Size	Trees Built	Time
con1	8/2	30	18	7
f0	8/1	24	23	7
f1	4/3	20	79	11
clp1	11/5	25	106	15
f2	4/4	38	302	33
dec1	4/7	67	450	58
z4	7/4	73	424	58
exam2	5/8	95	362	62
f3	8/7	96	378	62
f5	5/8	91	589	102
rd53	5/3	92	573	102
insdec	7/16	107	766	139
x1dn	27/6	143	758	146
vg2	25/8	145	868	153
5xp1	7/10	193	3136	565
dec2	8/7	202	3488	569
rd73	7/3	226	3643	635
9sym	9/1	351	2025	639
bw	5/28	271	7411	1003
x6dn	39/5	561	14371	3225
mlp4	8/8	504	17283	3295
dist	8/5	571	32512	4601
gary	15/11	741	32659	7443

system supports three target technologies.

The circuits synthesized by the SOCRATES system are comparable in cost to those produced by human engineers. A contest was held to compare the system to human experts. Two sets of minimized functions to be implemented, as well as a list of the types of gates available, were given to the participants. In order to keep the problems at a manageable level for the humans, no timing constraints were specified. Table 2 shows the results obtained by the experts and by SOCRATES; all sizes are in terms of transistor pairs. The experts spent between one and five hours on each example. A number of the contest entries contained logic errors that were quickly detected by using the extractor and comparator. In both cases SOCRATES generated more efficient implementations in very little CPU time. Figure 6 shows SOCRATES' implementation for the smaller of the two contest examples. While this design is very area-efficient, it is also quite slow. Figure 7 shows the circuit generated by SOCRATES when a constraint of seven nanoseconds was placed on the second output on the function. A third example (PLA1) in Table 2 was part of a standard cell design and had been optimized by a designer. An implementation generated by SOCRATES was used in the final circuit.

As the problem complexity increases, SOCRATES performs better comparatively, often performing better than humans, especially in cases with multiple outputs. With an improved knowledge base, the quality of the circuits clearly improves.

A rule-based expert system that optimizes combinational logic implementations in a given technology has been presented. The system allows the user to easily enter additional rules, which are automatically verified and classified in the knowledge base. Through the use of a set of metarules that manipulates the search control parameters (depth, breadth, etc.), the search space is kept small, resulting in

Table 2.
Comparison of SOCRATES to human experts.

Example	con1	con2	pla1
Human 1	42	81	—
Human 2	39	107	—
Human 3	36	—	—
Human 4	34	—	—
Human 5	—	—	165
SOCRATES	30 (7 sec)	70 (56 sec)	148 (143 sec)

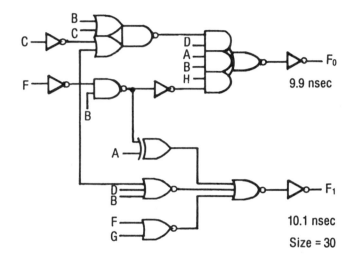

Figure 6. First contest example.

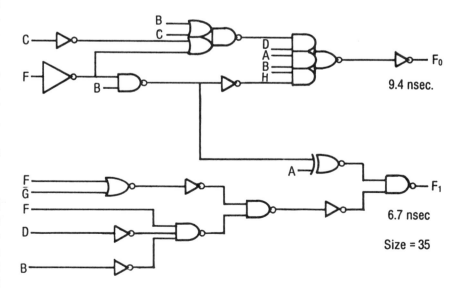

Figure 7. Second contest example.

short run times. The results obtained with the SOCRATES system are comparable in area and speed to circuits generated by experts. □

Acknowledgments

We would like to express our gratitude to the many people, and especially the summer and co-op students, who contributed to the ideas and implementation of this project. Special thanks to Karen Bartlett, University of Colorado; David Gregory, Stanford; Gary Hachtel, University of Colorado; Karl Garrison and Tony Hefner, North Carolina State University; and Bryan Kartzmann and Misha Rekhson, MIT.

References

1. R.K. Brayton, et al., *ESPRESSO-IIC: Logic Minimization Algorithms for VLSI Synthesis,* Kluwer Academic Publishers, Netherlands, 1984.

2. J. Darringer, et al., "Experiments in Logic Synthesis," *Proc. IEEE 'Int'l. Conf. Circuits and Computers,* 1980, pp. 234-237A.

3. J.A. Darringer, et al., "LSS: A System for Production Logic Synthesis," *IBM J. Research and Development,* Vol. 28, No. 5, Sept. 1984, pp. 537-545.

4. R.K. Brayton and C. McMullen, "The Decomposition and Factorization of Boolean Expressions," *Proc. Int'l. Symp. Circuits and Systems,* 1982, pp. 49-54.

5. D. Gregory, K. Bartlett, and A.J. de Geus, "Automatic Generation of Combinatorial Logic from a Functional Specification," *Proc. IEEE Int'l. Symp., Circuits and Systems,* May 1984, pp. 986-989.

6. J. McDermott, *R1: A Rule-Based Configurer of Computer Systems,* tech. rept., Dept. of Computer Science, Carnegie-Mellon University, Pittsburgh, Pa., 1980.

7. T.J. Kowalski and D.E. Thomas, "The VLSI Design Automation Assistant: An IBM System/370 Design," *IEEE Design and Test of Computers,* Vol. 1, No. 1, Feb. 1984, pp. 60-69.

8. W.W. Cohen, K. Bartlett, and A.J. de Geus, "Impact of Metarules in a Rule Based Expert System for Gate Level Optimization," *Proc. IEEE Int'l. Symp. on Circuits and Systems,* May 1985.

9. R.B. Hitchcock, "Timing Verification and the Timing Analysis Program," *Proc. IEEE/ACM 19th Design Automation Conf.,* 1982, pp. 594-604.

10. C.L. Forgy, *OPS5 User's Manual,* technical report, Dept. of Computer Science, Carnegie-Mellon University, Pittsburgh, Pa., 1981.

11. W.F. Clocksin and C.S. Mellish, *Programming in Prolog,* Springer-Verlag, New York, NY, 1981.

The authors' address is General Electric Microelectronics Center, Semiconductor Business Division, PO Box 13049, Research Triangle Park, NC 27709.

Synthesis and Optimization of Multilevel Logic under Timing Constraints

KAREN BARTLETT, WILLIAM COHEN, AART DE GEUS, MEMBER, IEEE,
AND GARY HACHTEL, FELLOW, IEEE

Abstract—The automation of the synthesis and optimization of combinational logic can result in savings in design time, significant improvements of the circuitry, and guarantee functional correctness. Synthesis quality is often measured in terms of the area of the circuit on the chip, which fails to take into account the timing constraints that might be imposed on the logic. This paper describes SOCRATES, a synthesis system capable of generating combinational logic in a given technology under user-defined timing constraints. We believe this system is the first to perform optimized, delay-constrained, multilevel synthesis into standard cell libraries. Applied to a large number of examples, the system has successfully traded off area versus delay and performs optimized, delay-constrained, multilevel synthesis into standard cell libraries.

I. Introduction

THE AUTOMATION OF the synthesis and optimization of combinational logic can result in savings in design time, significant improvements of the circuitry, and guarantee functional correctness. Synthesis quality is often measured in terms of the area of the circuit on the chip, which fails to take into account the timing constraints that might be imposed on the logic.

This paper describes SOCRATES, a synthesis system capable of generating combinational logic in a given technology under user-defined timing constraints. The user-defined timing constraints are used at two places in our system: during the multilevel function synthesis which yields the structure of the circuit and during the gate level optimization of the circuit which yields the technology-dependent implementation of the logic.

The synthesis of combinational logic can be performed either at the transistor level or using a predefined set of logic gates. PLA's fall in the first category and their automatic generation and area optimization have been studied extensively [4], [11]–[13]. The PLA is a two-level structure with limited potential for timing optimization. Recently, methodologies for multilevel implementation at the transistor level have been proposed based on the use of CMOS domino logic. One of these transistor-level implementations is employed by the "YLE" synthesis sys-

tem [5] and uses a standard cell-like "pluricell" methodology. The other uses a "PLA-like" layout scheme [14], but both of these transistor-level systems require specialized layout generation programs.

Significant previous work has been done on optimized synthesis into gate array libraries, e.g., MACDAS [16] and LSS [9]. The YLE system [6] performs optimal **multilevel** synthesis using "WEAK DIVISION" into domino pluricells, but is not library based. DeGeus *et al.* [8], [10] have reported on an expert systems approach which uses an optimized sequence of local transformations on a multilevel system produced by WEAK DIVISION. However, in all of these previous investigations, little work has been done to automatically generate optimized, multilevel library-based combinational circuitry while simultaneously meeting timing constraints.

The approach presented in this paper builds on all the referenced approaches. It is an extension and elaboration of the work described in [1] and [2]. The synthesis is performed in two main phases: 1) algorithmic creation of the "logic structure" of the circuit, which is then "mapped" into a given logic gate library, and 2) a set of local transformations translating and optimizing the circuit in the target technology, which underlies the given library. In the first phase, a multilevel circuit is built using a variant of WEAK DIVISION. The timing constraints influence the WEAK DIVISION process through the use of approximate delay models and thus shape the "architecture" of the circuit in terms of its delay–area tradeoff. In the second phase, optimization is performed by a rule-based subsystem, called OPTIMIZE, in which the implementation technology is fully known and an exact delay computation is available. A set of timing-specific rules transforms the circuit trading off area versus delay when necessary to meet the timing constraints imposed by the designer.

We make and utilize throughout this paper the following "cooperation assumption":

WEAK DIVISION and OPTIMIZE are separate and distinct circuit optimization processes, in which the output of the WEAK DIVISION process is the input to OPTIMIZE. We assume that the quality of the OPTIMIZE output improves with the quality of the WEAK DIVISION output. In most cases, with rare exceptions noted in Section V, this assumption appears to be justified. In an important sense, as discussed in Section V below, if either of these two

Manuscript received January 22, 1986; revised May 23, 1986. This work was supported in part by the National Science Foundation under Grant ECS-8121446.

K. Bartlett and G. Hachtel are with the Department of Electrical and Computer Engineering, University of Colorado at Boulder.

W. Cohen and A. De Geus are with the General Electric Microelectronics Center, Research Triangle Park, NC 27709.

IEEE Log Number 8609878.

Reprinted from *IEEE Trans. CAD of Int. Circ. Syst.*, vol. CAD-5, no. 4, pp. 582–596, Oct. 1986.

modules were perfect, there would be no need for the other.

The sequel begins in Section II with an overview which provides a brief description of each of the system components and how they interrelate. Section III describes WEAK DIVISION and MULTILEVEL MINIMIZATION, the techniques used to obtain an efficient multilevel representation of the function. Section III also describes the cost and delay models used in multilevel synthesis. Section IV describes the module performing local transformations, which is a rule-based expert system. Results and conclusions are covered in Section V and VI.

II. OVERVIEW

Fig. 1 shows the architecture of the SOCRATES logic synthesis system. The application-specific input to the system is a functional specification which can take one of three different forms:

1) a set of Boolean equations, possibly multilevel;
2) a set of "linked" PLA's (single output) in ESPRESSO format;
3) a "net list" of library gates.

Each of these forms can be represented graphically by a so-called "Boolean network," as illustrated in Fig. 2. Each node in a Boolean network is a two-level function in one of the three forms listed above. The Boolean network is a directed acyclic graph, whose edges represent the logic dependencies implied by any of the three input forms. Note that nodes without "fan-in" are primary inputs (pentagons in Fig. 2).

As shown at the bottom of Fig. 1, the output of the system is a net list of technology and application-specific library gates. This output is dependent on additional input such as the gate library, the rules library, and the specification of timing and fan-in constraints, which are unrelated to the application. Further, design guidelines, such as the type of delay models to use, can also be considered as system inputs.

The problem of synthesis under timing constraints is viewed as a set of translation and optimization problems, where each optimization is carried out at a different level of abstraction. The first level of abstraction is the sum-of-products level. The process of simplifying the two-level Boolean equations at each node of the given Boolean network is called *minimization* and is carried out by the ESPRESSO IIC logic minimizer [4], [15]. Synthesis at the second level of abstraction involves the creation of an optimum multilevel Boolean network, which is the output of the SYNTHESIS module (dashed box in Figure 1). This module is comprised of the WEAK DIVISION, MULTILEVEL MINIMIZATION, and LIBRARY MAPPING submodules.

The technique of *weak division* [3], [6] is used to decompose the given Boolean network, which may or may not have more than two levels, into an alternative optimized multilevel Boolean network. This "structurally op-

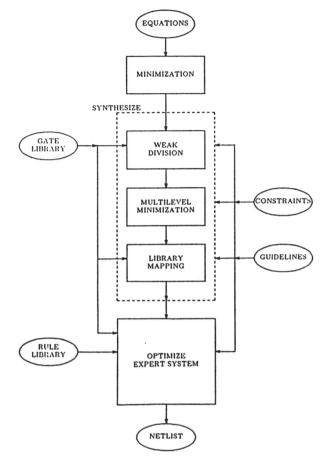

Fig. 1. SOCRATES system overview.

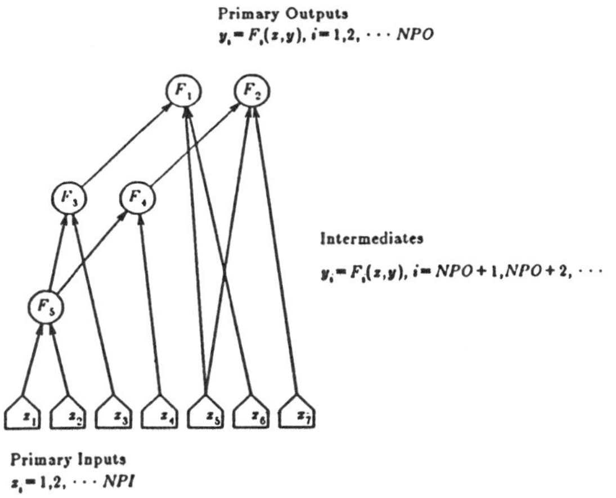

Fig. 2. Multilevel Boolean network.

timized" Boolean network begins to reflect the structure of the final circuit; it is apparent at this stage what logic will be re-used, and approximately how many levels of logic will be used to implement the function. Synthesis of this multilevel Boolean network is heavily influenced by timing considerations. The timing delay models used are discussed in Section III.

After a decomposition is found that is satisfactory from a performance and area standpoint, *multilevel minimization* [6] is used to simplify this decomposition. This technique is based on the "don't care" conditions associated with the intermediate variables introduced by the decomposition into a multilevel function. Transformations suggested by this process are rejected if they do not significantly reduce area or if they increase the estimated delay.

The function(s) is then translated into a logic circuit by replacing the two-level logic functions associated with each node of the Boolean network into library specific logic gates. This mapping into the library may use generic "dummy" gates, like AND's and OR's, which are, ultimately, replaced by the OPTIMIZE module with actual gates from a user-supplied library.

Finally, the circuit implementation is optimized in the target technology. A rule-based system, OPTIMIZE, performs local transformations formulated as rules on the circuit. The optimality of the final circuit thus produced depends on the rules in the library and the order in which these rules are applied. Our approach uses a state space look-ahead algorithm to optimize the application order, as discussed in Section V (logic level synthesis and optimization).

The optimization criterion is based on both circuit area and circuit delay. The system will try to meet the timing constraints and subsequently perform area optimization. The timing constraints will guide the specifics of the implementation leading to a locally optimal choice of gates in the given technology.

If the end product does not satisfy the user, or if the user wishes to consider a number of different possible designs, then the timing constraints can be changed and the procedure iterated.

It is appropriate to think of the SYNTHESIS and OPTIMIZE modules of Fig. 1 as interdependent, interacting, heuristic optimization processes.

III. MULTILEVEL SYNTHESIS AND MINIMIZATION

In this section, we will describe the WEAK DIVISION and MULTILEVEL MINIMIZATION portions of the SYNTHESIS module. Together, these two algorithms can be regarded as a program for finding an optimum architectural restructuring of the given, initial, often two-level, Boolean network. In particular, we will describe how this program acts to minimize the cost function of Figure 3, which, roughly speaking, serves to minimize the gate array (or standard cell) area subject to user-specified delay guidelines.

The plan of the section will be as follows. First, we describe, in Section III-A, the computation of network "cost" as required by WEAK DIVISION and MULTILEVEL MINIMIZATION. Because the cost constraints of Figure 3 depend on critical path delay, we will also discuss the two basic delay models employed in the SYNTHESIS module:

1) the "unit delay" model (which is technology- and fan-out-independent);
2) the "library element gate delay" model which employs the basic delay equation of Figure 4.

$$
\begin{aligned}
p &= \text{Primary Inputs} \\
b(p) &= \text{Given Boolean Function} \\
\eta &= \text{Initial Boolean Network} \\
&\quad \text{Realizing } b(p) \\
N^{\bullet} &= \text{Set of all Boolean Networks } \eta \\
&\quad \text{Realizing } b(p) \\
\tau(\eta) &= \text{Critical Path Delay} \\
g &= \text{Logic Gate} \\
\text{Area } (\eta) &= \sum A(g) \\
\text{Wires } (\eta) &= \sum_{g \in \eta}^{g \in \eta} |\text{ FAN\_IN } (g)|
\end{aligned}
$$

$$
\text{Objective: } \eta^{\bullet} = \underset{\eta \in N^{\bullet}}{\text{ARGMIN}} \{\text{AREA}(\eta) + \text{WIRES}(\eta)\}
$$

$$
\text{Delay Constraint: } \tau(\eta) \leq \text{CHECK\_DELAY}(\eta^0, \text{guidelines})
$$

Fig. 3. Delay constrained area optimization via decomposition.

- Levels of Logic Mode (each gate has "unit" delay)
  - "Architectural" Level Approximation
  - Easy Critical Path Problem
- Library-Element Gate Delay Mode
  - Fanout Dependent
  - $C_{gp'}$ = Capacitance of Input Pin p' of Gate g
  - RLOAD_g = Output Load Resistance of Gate g
  - TSET_g = Intrinsic Delay of Gate g

$$
\tau_g = \text{TSET}_g + \text{RLOAD}_g \cdot \sum_{p' \in \text{FANOUT}(g)} C_{gp'}
$$

Fig. 4. SYNTHESIS delay models.

Once the cost and constraint computations are defined, we can state the problem to be solved by the SYNTHESIS module as follows:

Create, in terms of library elements, an optimally structured circuit which, when subsequently processed by the OPTIMIZE module, produces the best final implementable Boolean network.

The basic "cooperation assumption" is that the better the SYNTHESIS module does in solving its optimization problem, the better the OPTIMIZE module will do in solving its problem (a similar optimization with a similar cost function). (The validity of this assumption is tested in Section V below.)

In Section III-B, we describe the WEAK DIVISION optimization process and in Section III-C, we discuss overall program control of this process. In Section III-D, we conclude by describing, very briefly, the MULTILEVEL MINIMIZATION process and presenting some limited experimental results.

A. Cost and Delay Estimation

Decisions about what synthesis and optimization steps to take requires the ability to obtain cost and delay estimates for all subfunctions which comprise the multilevel multiple output function.

Cost and delay estimates are obtained by doing a

straightforward mapping on each subfunction in order to generate a gate level representation in terms of the primitives in the user-specified gate library.

The gate library contains two types of elements: implementation primitives and generic elements. The following information is required for each element in the library:

a) the Boolean function defining the cell;
b) the cell area (the number of "transistor pair" units);
c) the terminal count of the cell;
d) delay information for the cell;
e) the equivalent output (load) resistance of the gate;
f) the pin capacitance of each gate input.

Implementation primitives correspond to primitives in the gate library such as NAND2, NAND3, and AOI33. From this library, actual area and delay attributes can be obtained for each element. Generic primitives are ANDS, ORS, and sums of product expressions satisfying fan-in constraints of the implementation primitives. The generic primitives are interim, "dummy" gates which will always be deleted by the OPTIMIZE module discussed in Section IV.

Given a gate level representation, the "AREA" and "WIRES" terms of the cost function in Figure 3 are computed by summing the gate area and terminal count for all gates in the Boolean network. The terminal count serves to estimate the amount of routing that will be needed between gates.

The delay constraint term $\tau(\eta)$ is obtained by performing critical path analysis on the current Boolean network. The delay information needed depends on the level of delay modeling used. When operating in "unit delay" mode, only the set delay through each gate is used. If the "library element gate delay" mode, which incorporates fan-out, is used, it is also necessary to provide the output pin resistances and input pin capacitances.

During straightforward mapping, if the representation of a function or candidate subexpression is in one-to-one correspondence with a primitive in the library, the library-specified area, interconnect, and delay numbers associated with the primitive are used. If there is not a direct mapping, the function is broken down and constructed from a set of generic primitives. Each product or sum is implemented in a few gates as possible without violating fan-in or fan-out constraints. Delay through the configuration is kept down by ordering inputs to each AND or OR representation by their propagation delay. The inputs with the longest delay are assigned to the inputs in the representation with the most "slack" (cf., discussion of (1) below).

The library elements used to implement an intermediate function may implement either the function or its complement. Determining which form of the function is best to implement is known as phase assignment. The best phase for a function depends on the phases available for its inputs and the phases needed by its fan-outs. The overall optimum phase assignment problem is NP-complete. Because of this, neither the mapping nor the cost and delay routines are concerned with determining the best phase

for a function. Both phases of all primary inputs and intermediate variables are assumed to be available. We have been investigating optimal phase assignment by simulated annealing, and plan to report our results in a subsequent paper.

The SOCRATES system considers Boolean networks with multiple primary outputs with specified target delays for each output. For simplicity in discussing delays, we shall assume that each Boolean network has a single primary output. The propagation delay through any given function (node of the Boolean network) is the time it takes for a signal to propagate from the primary inputs to the output of the given function. This delay depends on the local propagation time through the gate(s) implementing the function and the propagation delay of the function's inputs. If the implementation of the given function representation requires multiple library gates, their configuration will be based on the arrival time of the functions inputs in order to reduce the total propagation delay. The delay at the output of a given node f is

$$\tau_f = \text{TSET}_f + \max_{i \in \text{FANIN}(f)} (\tau_i - \text{local\_slack}(f, i)). \quad (1)$$

TSET_f refers to the delay through the gates which implement this function. When operating in unit delay mode, this would be one unit for each level of logic implementing the gate. In library element gate delay mode, the gate delay equation shown below is used:

$$\tau_g = \text{TSET}_g + \text{RLOAD}_g^* \sum_{p \in \text{FANOUT}(g)} C_p \quad (2)$$

where C_p is the input capacitance associated with the pin to which g fans out. When the mapping requires multiple gates, the set delay is the sum of the gate delays and load resistance components of the longest path through the configuration. The load resistance is the load resistance of the final gate, i.e., the gate whose output is the value of f. "local_slack"(g, i)) refers to the time after start of computation for f that i can arrive before it is considered critical, and is a function of the configuration. It is computed by ordering the inputs by their delays ($O(n \lg n)$, heapsort); those with the most delay are assigned to the input pins with the most local_slack. Also, the total input capacitance of function f fanning out to a function which is realized with multiple gates depends on the number of gates to which f fans out.

The user may specify propagation delays for primary inputs reflecting their relative arrival times. If delays are not specified, they are assumed to be 0.

Prior to computing the delay through each function, the graph is levelized based on function dependencies. A function which depends only on primary inputs is a level 1 function. A function depending on a level i function is a level $i + 1$ function. The delays for the primary inputs are read in or assigned default values of 0. The remaining delays are then computed by levels, starting with level 1.

When the delay of a function changes, it may affect the delay of those functions in its (transitive) fan-out. Changes are propagated one level at a time until either a primary output has been reached or the delay of a function does not change. The local slack for each function needs to be recomputed as different delays or inputs may result in a different assignment of inputs to pins.

If the library gate delay mode is used, then a "latent" critical path delay calculation is performed to propagate a local delay change, as calculated by (2) through to the primary output. Here we simplify the calculation with the following assumption:

Assumption: Only the (transitive) fan-out of a gate g in a given Boolean network is affected by a change in the delay of a transformed gate.

This assumption actually constitutes an approximation in the library element gate delay mode because the substitutions made by WEAK DIVISION affect the fan-out of gates not in the (transitive) fan-out of the given gate g. Hence, by the calculation of Fig. 4 the delay through these other gates also could, in principle, change. These changes are ignored when computing the cost of candidate substitutions in order to reduce the overall running time of WEAK DIVISION. Of course, once the best candidate has been selected and substituted, an exact critical path delay calculation is made. This latter calculation is still latent, however, in the sense that it startes only at nodes which fan-in to nodes whose delay has actually changed since the last transformation (i.e., substitution).

B. Library-Based WEAK DIVISION

As discussed above, algebraic decomposition or WEAK DIVISION is a method of recognizing the subexpressions which are either common to two or more different functions or factorize, and thus simplify, one individual function.

Example (Weak Division on Two-Level Function):

Initial representation	Modified representation
F_1: $aef + bef + cef$	F_1: Bef
F_2: $aeg + beg + deg$	F_2: Ceg
	A: $a + b$
	B: $A + c$
	C: $A + d$

If $A = F_5$, $B = F_4$, $C = F_3$, and $(a, b, \cdots, g) = (x_1, x_2, \cdots, x_7)$, the Boolean network which corresponds to the modified representation is that of Fig. 2. In the sequel, we shall refer to product terms such as *aef* as **cubes**. Hence, each Boolean function is regarded as a set of cubes. The true or complement form of a Boolean variable $(a, \bar{a}, b, \bar{b} \cdots)$ is called a **literal**. In the above example, the primary output functions F_1 and F_2 were reexpressed by detecting and creating intermediate variables for common subexpression A and factors B and C, and substituting these into the original representations. The functions in the initial two-level Boolean network de-

```
Procedure WEAK_DIVISION
Begin

    /* DECOMPOSITION */
    While (common subexpressions exist)
        Generate candidate subexpressions for
            current functions
        Determine eligible subexpressions
        Select "best" disjoint subexpressions
        Associate new intermediate variables with
            subexpressions and substitute
    Endwhile
    Collapse subexpressions referenced by
        only one function

    For each function (FACTORIZATION)
        repeat above loop for single function
    Endfor
End WEAK_DIVISION
```

Fig. 5. WEAK DIVISION algorithm.

pend only on primary inputs. The functions in the multilevel "modified" Boolean network (four-level in this case) are expressed in terms of both primary inputs and intermediate variables. The SYNTHESIS module permits the initial representation to be either two-level or multilevel.

As suggested by the above example, the WEAK DIVISION process regards the two-level functions associated with the nodes of the Boolean network as algebraic expressions. The process consists of a set of decomposition steps, each one involving the following substeps:

a) determination and examination of candidate subexpressions;
b) selection of the "best" candidate subexpressions;
c) substitution of best candidate subexpression (which transforms the current Boolean network into an altered one).

This process serves two purposes. First, it enables sharing of logic between the multiple functions. Second, it is the basis for decomposition of the original representation into primitives which are more readily implementable in the desired target technology.

The algorithm for WEAK DIVISION, outlined in Fig. 5 shows that the WEAK DIVISION process is an iterative one. Each iteration has four phases: the generation of candidate subexpressions, the pruning of subexpressions which do not satisfy user constraints, the selection of the best disjoint subexpressions, and the substitution of these subexpressions into one or more of the functions which they divide. The substitution of a subexpression into one or more functions may create new divisors which are expressed in terms of this new intermediate variable. This process continues until there are no subexpressions of sufficient merit to warrant further substitution. WEAK DIVISION is separated into two steps: *decomposition*, the recognition of subexpressions common to multiple functions, and the *factorization* of individual functions. Any intermediate variables which are only referenced by a single function are collapsed back into the function between these steps. If there are still functions which are not read-

```
Procedure KERNEL_GEN (F,start_lit)
    If CUBE_FREE(F) Then
        RECORD_KERNEL(F,LEVEL(F))
        If (LEVEL(F) == 0) Then
            Return
        Endif
    Endif
    For lit←start_lit,last_lit
        If (lit in F)
            subf←F/lit
            subf←subf/COMMON_CUBE(subf)
            KERNEL_GEN(subf,lit + 1)
        Endif
    Endfor
    Return
End KERNEL_GEN
```

Fig. 6. The Kernel construction algorithm.

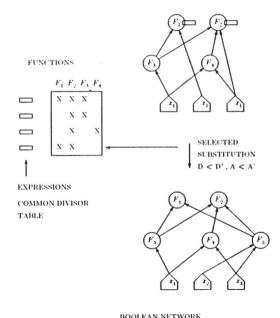

FUNCTIONS

	F_1 F_2 F_3 F_4
▭	X X X
▭	X X
▭	X X
▭	X X

SELECTED SUBSTITUTION
D < D', A < A'

EXPRESSIONS

COMMON DIVISOR TABLE

BOOLEAN NETWORK

Fig. 7. WEAK DIVISION process.

ily represented in the primitives of the target technology, these are broken down in a straightforward manner.

Algorithms based on those presented in [3] are used to generate to candidate subexpressions and to substitute the selected subexpressions. The algebraic cofactor h of f with respect to g ($h = f/g$) is a new function defined to be the largest set of cubes (product terms) such that h and g have no literals in common and every term in the Boolean intersection of h and g (hg) is in f, i.e., $f = (f/g)g + r_g = hg + r_g$, where r_g is the appropriate remainder. In the example above, if $g = A = a + b$, then $F_1/g = ef$, and $r_g = cef$. Note if $f/g \neq \phi$, then both g and f/g are divisors of f.

During the decomposition phase, there are two types of candidate subexpressions, those generated by *distillation* and those generated by *condensation*. There are parameters to guide which type to generate during each pass. Options exist to generate both or to generate distillation terms until the merit drops below a user-specified value and then to generate both.

Candidate distillation expressions are subexpressions of **kernels** which appear in more than one function. The **kernels** K of an expression f are formally defined as a set of cubes $K(f)$ which satisfies

$$K(f) = \{f/c: c \text{ is a cube} \quad \text{and} \quad f/c \text{ is cube free}\}. \quad (3)$$

An expression is **cube free** if there is no literal which appears in every cube. Note that a single product term is not cube free; thus, a kernel is defined to contain two or more product terms.

In the Example, the only kernel of F_1 is $a + b + c$, which is F_1/ef. F_1/e is not cube free and F_1/a, F_1/b, F_1/c do not contain two or more cubes.

The kernel construction algorithm shows in Fig. 6 is a recursive algorithm. Initially, F is the function to obtain kernels for and *start_lit* is set to 0. A level 0 kernel K is defined to be one where $K(K) = K$. A level $n + 1$ kernel is one where $K(K)$ are all of level $\leq n$. In a level 0 kernel, no literal appears in more than one cube. The kernels for each function are generated independently, but all kernels are stored in a common kernel table whose rows correspond to the kernels and whose columns correspond to the functions divided by the kernel.

After generating kernels for all functions, additional common divisors are found and added to the kernel table. A common divisor is any $K_i \cap K_j$ such that K_i is a kernel from function i, K_j is a kernel from function j, and there are at least two cubes which appear in both kernels. Associated with each divisor are the cubes comprising the divisor and the functions which it divides.

The complement of each kernel is also generated and each function is checked to see if it can be divided by the complement. Those kernels which do not divide multiple functions will not be considered for substitution. The above process is illustrated in Fig. 7. At the top right is shown the Boolean network prior to the selection and substitution of a common divisor. At the left, a table is shown where the rows correspond to common divisors and the columns to functions (nodes of the Boolean network). The entries X may take one of three values, indicating the "signature" of the expression, i.e., whether the division function occurs in true form, complement form, or in both forms in the function associated with its column. No entry means the divisor does not occur in that function (column). Divisors not common to two or more functions are not present in the table. The cost is computed, as described in Section III-A, for each candidate substitution, and the resulting transformation of least cost is selected.

Note that, generally speaking, divisors which are common to more functions provide greater area savings. However, in the case illustrated in Fig. 7, a divisor common to only two functions (F_1 and F_2) is selected, despite the existence of a divisor (the first) common to three functions (F_1, F_2, and F_3). This is because of the connectivity of the Boolean network. Substitution of the latter divisor would result in an extra level of logic. This subexpression would not be considered if it resulted in violating any delay constraints.

Candidate condensation subexpressions are cubes which can cofactor multiple functions. A condensation table is

39

constructed containing all such cubes. Associated with each are the literals comprising the cube and the functions which it can cofactor.

A function f can be reexpressed in terms of a subexpression g if f/g is not empty. The complement of the candidate subexpression is also considered for substitution. Thus, as a result of substitution, we have a new representation of the function f in the form $f = (f/g) G + (f/\overline{g}) \cdot \overline{G} + r_g$, where g is the expression to be substituted and G is the literal associated with it. As a subexpression may have been substituted into other functions on a previous pass, it is necessary to check if it has already been implemented. If neither the subexpression or its complement exist in the function table, the subexpression is added and associated with a unique literal. All eligible occurrences of the subexpression and its complement are then replaced with the corresponding literal or its complement. Two subexpressions g and h are disjoint if $(f/g) g$ and $(f/h) h$ do not contain any of the same cubes or if $f/g \supseteq h$ and $f/h \supseteq g$.

As the common subexpressions are typically quite small, the number of kernels to consider can be reduced greatly by considering only level 0 kernels. This also speeds up the kernel intersection process. Often, the more complicated common divisors can be obtained by back substituting (collapsing) any literals referenced by a single intermediate variable.

The algorithms used for factorization are very similar to those used for decomposition. Only one function is considered at a time and the candidate subexpressions are the level 0 kernels of each function.

C. Program Control of WEAK DIVISION

The objective of our WEAK DIVISION process is to select those subexpressions that will yield the best implementation in the specified target technology. After each generate phase, merit and delay increment estimates are obtained for the candidate subexpressions. The best disjoint subexpressions become intermediate functions. Different candidate subexpressions will generate alternate representations of the functions into which they can be substituted. Selecting one intermediate subexpression may eliminate other candidate subexpressions.

The user may influence the final propagation delay by specifying delay constraints. These prevent, regardless of merit, the substitution of any candidate subexpressions which would increase the delay of any critical primary output functions by more than a user-specified amount. There are several different user-specified ways a primary output function can be considered critical:

   a) exceeds a certain absolute delay figure;
   b) exceeds the maximum initial delay;
   c) exceeds the maximum current delay.

Given a criteria for determining if a primary output function is critical, the slack associated with each primary output is calculated as follows. If the initial network is multilevel or if kernels of level > 0 are allowed, it is necessary

to determine how much a change in the delay of an intermediate function will affect the delay of the primary output functions in its transitive fan-out. This can be done by maintaining an array showing the longest path between each primary output and each intermediate function. The slack σ_j associated with each intermediate function F_j is

$$\sigma_j = \underset{i \in PO}{\text{MIN}} \{\sigma_i + P(i) - (P(j) + LP(i, j))\} \quad (4)$$

where PO denotes the set of primary output indices in the transitive fan-out of j, $P(j)$ is the length of the longest path from any primary input to the intermediate function F_j, and $LP(i, j)$ is the length of the longest path to primary output function F_i from intermediate function F_j.

If substituting a candidate subexpression into a function causes a delay increase which is greater than the slack calculated for the function, then the subexpression is no longer considered for substitution in this function.

The merit $M(G, F)$ of substituting a subexpression G into a specific function F is the difference in cost between the original (f) and the revised representation $(f' = (f/g) G + (f/\overline{g}) \overline{G} + r_g)$, i.e.,

$$M(G, F) = \text{COST}(f) - \text{COST}(f'). \quad (5)$$

If the subexpression G is to be substituted into the functions F_j, $j \in J_G$, where J_G is the set of functions G can cofactor without violating delay constraints, then the total merit $MT(G)$ is given by

$$MT(G) = \left(\sum_{j \in J_G} M(G, F_j)\right) - \text{COST}(G) \quad (6)$$

where $\text{COST}(G)$ is as defined in Section III-A.

D. Multilevel Function Minimization

As WEAK DIVISION is a strictly algebraic process, the network may contain intermediate or primary output functions that are functionally equivalent. Because of "don't care" conditions inherent to multilevel Boolean networks, functions which are prime and irredundant in the two-level sense may be nonminimal in the multilevel sense. As a result, the network may also contain functions which can be more efficiently expressed in terms of a different set of intermediate functions. To exploit these possibilities, we now discuss a technique we call "MULTILEVEL MINIMIZATION."

The "two-level" notions of primality and irredundancy are yet to be rigorously defined in the multilevel context, but the don't care methods described below establish a promising multilevel minimization technique for two-level minimization. In fact, if the full don't care set described below is used, then there is a meaningful sense in which the resulting network may be called prime and irredundant. We plan to explore these matters more fully in a later paper [17]. For now we content ourselves by thinking of MULTILEVEL MINIMIZATION as a process which, like WEAK DIVISION, reduces the cost function of Fig. 3 by a series of local transformations. In this case, the transformations are those made by the ESPRESSO logic minimizer given the multilevel don't care set discussed below.

In two-level minimization, we are given $F: B^n \to B$ and attempt to minimize $\|F \cup D_x\|$, where $\|\cdot\|$ is some measure of term count (circuits) and literal count (transistors) in the given cubical cover of F. Here, $D^x = \{x \in B^n| \text{ "}x \text{ never occurs as input"}\}$ is the so-called don't-care set [4].

In multilevel minimization, we are given a set of primary inputs x_i, $i = 1, 2, \ldots, n$, and a set of intermediate variables $y_k = F_k(x, y)$ $k = 1, 2, \ldots, m$. Note that here $F_k: B^{m+n} \to B$. The y_k can be primary outputs or literals representing intermediate variables which are output from node k of the Boolean network example of Fig. 1.

The "intermediate" don't care set associated with function F_j of the Boolean network is given by

$$D_j^I \equiv x_j \overline{F}_j(x, y) + \overline{x}_j(x, y) . \qquad (7a)$$

Simply stated, D_j^I gives the set of vertices of B^{n+m} which have inconsistent combinations of the components of x and y vectors, i.e., combinations which cannot occur, and are, therefore, "don't care." The totality of such intermediate don't care conditions

$$D^I = \sum_{j \in 1, 2, \ldots, m} D_j^I \qquad (7b)$$

is known to be potentially prohibitively large and, when minimizing function F_k of the Boolean network, is usually replaced by a subset D^{Ik}, defined by

$$D^{Ik} = \sum_{j \in jk} \sum_{i \in Ij} D_i^I \subseteq D^I \qquad (7c)$$

where I_j is a set of node indices and J_k is a set of set indices. The purpose of the double summation is to permit sufficiently powerful subsets of D^I to be expressed which, it is hoped, are minimally sufficient for the complete minimization of F_k. We are currently using the transitive fan-out of the transitive fan-in of node k, minus the transitive fan-out of node k for that purpose.

In terms of these definitions, the multilevel don't care set for function F_k can be expressed as follows:

$$D_k = D^z \cup D^{Ik} \cup D^{Ok} \qquad (7d)$$

where D^{Ok} stands for the "output" don't care set [7], which we have neglected in the work described below. The effects of D^{Ok} will be discussed in a later paper [17]. If D^{Ok} is neglected, only primary output functions will necessarily be rendered prime and irredundant by our procedure. However, any of the F_k may possibly be simplified, thus reducing the cost function of Figure 3, whether or not they are primary outputs.

Assuming that $D^{Ik} \equiv D^I$ and that node k of the Boolean network is a primary output, for which $D^{Ok} = \Phi$, then it can be shown that (7d) gives the complete don't care set for function F_k. If a complete don't care set is used, for each F_k, then after ESPRESSO is applied to function $F_k \cup D_k$, for all $k = 1, 2, \ldots, m$, the resulting Boolean network can be shown to be prime and irredundant in the sense that no cube or product term of any functions can

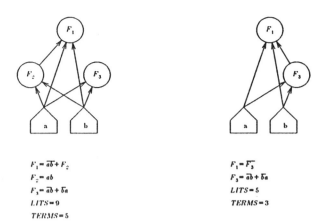

$$F_1 = \overline{a}b + F_2$$
$$F_2 = ab$$
$$F_3 = \overline{a}b + \overline{b}a$$
$$LITS = 9$$
$$TERMS = 5$$

$$F_1 = \overline{F_3}$$
$$F_3 = \overline{a}b + \overline{b}a$$
$$LITS = 5$$
$$TERMS = 3$$

Fig. 8. MULTILEVEL MINIMIZATION example.

be deleted without altering the Boolean functionality of the Boolean network.

However, with any don't care set which is complete in the sense described above, the adjacency relations of the Boolean network may be altered as shown in the example of Fig. 8. It is of interest to observe that in this example the starting representation (on the left) is prime and irredundant. Thus, the first EXPAND and IRREDUNDANT-_COVER operations in ESPRESSO will have no effect. However, after the REDUCE operation is performed, the optimal result will be obtained in the second EXPAND step. We observe, in fact, that REDUCE is performing a major part of the role of the minimization process referred to as Boolean substitution in [6]. This effect offers a major avenue for future research.

IV. LOGIC-LEVEL SYNTHESIS AND OPTIMIZATION

The final step in the synthesis of a circuit is to implement the optimal multilevel function at the logic level. The challenge here is to make the best possible use of the target technology. The approach taken to this problem is to first map the multilevel function to a logic-level circuit, and then to optimize the synthesized circuit. Since this module of the synthesis system has been described in earlier publications [10], we will merely summarize in this paper its workings and concentrate on the timing specific parts.

A. Initial Mapping to Target Technology

The translation of a multilevel function to a circuit is a straightforward operation. In order to make this translation tool technology-independent, the initial circuit is always implemented using the same small subset of library gates. If these gates do not actually exist in the target technology, then a high cost is assigned to them to force their later removal.

B. The SOCRATES Rule-Based System

The optimization of the circuit at the gate level is done by a rule-based system. A knowledge-based approach, rather than an algorithmic approach, was used in order to provide a paradigm for the use of specialized knowledge about the target technology. A circuit is optimized by performing *local transformations* on it: each transformation replaces one small configuration of gates with another configuration that is functionally equivalent. A cost function similar to that described in Section III is used to assess the merit of the replacement. Any given transformation may reduce the cost of the circuit from the standpoint of either area, performance (i.e., delay), or both.

Knowledge acquisition has traditionally been difficult in existing rule-based systems. A *rule entry module* aids the knowledge engineer in adding new local transformations to the system. This module automatically verifies all new rules for functional correctness and orders them in the knowledge base. The ordering is determined by the area and delay benefits associated with the rule in question and is also a function of the relationship of the transformation to the rules already in the knowledge base.

During the optimization, the system performs four sets of tasks.

1) A pattern matcher is used to find which rules apply on the circuit. Generally, more than one rule can be applied at any time during the optimization.

2) For each potential rule application, a cost function is computed to determine the quality of the resulting circuit.

3) A selection mechanism decides which rule is to be applied next.

4) The rule is applied by constructing the new gate configuration on top of the old configuration, removing the old one and adjusting the cost function to its new value.

C. The Search Strategy

The actual sequence of transformations that will be used is determined by a control module which "looks ahead" several rule applications in order to select the most useful set of transformations. This is done in order to avoid local minima of the cost function. Although the amount of CPU time required increases rapidly with increased breadth and width of the search space explored for each move, it was found that on the average the resulting circuit size can be reduced by an additional 10 percent using a look-ahead strategy. The search space explored is the space of Boolean networks which can be obtained from the initial library mapping by some sequence of rule applications. If the Boolean network has N_V nodes and there are N_R rules in the rules library, there are, potentially

$$N_V \sum_{k=1}^{N_R} k!$$

such sequences. We explore a meaningful subspace of this potentially large state space by a search algorithm that is illustrated in Fig. 9. This algorithm is controlled by a number of parameters, the most important of which are breadth (the number of rule sequences explored from a given network configuration) and depth (the length of the rule sequence to be explored).

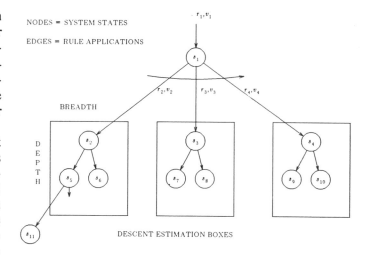

NODES = SYSTEM STATES

EDGES = RULE APPLICATIONS

SEARCH METHOD

1. LOOK AT ALL STATES IN EACH SON'S ESTIMATION BOX

2. TAKE BEST RULE (r_2) TO NEW STATE (s_2) AND RECUR

3. IF BOX CONTAINS SUFFICIENTLY MERITORIOUS STATE

4. THEN SEARCH RECURS FROM s_5

Fig. 9. State space search.

D. Meta-Rules

During the development of the optimizer, it became apparent that the search parameters should be dynamically adjusted during optimization to improve results and run time. A second rule-based system was therefore built to control the search strategy of the system. A set of *meta-rules* modifies the control parameters of the search mechanism based on a set of diagnostics from the system. While the transformation rules used reflect the expertise of the circuit designer, the meta-rules capture the experience of the designer of the optimization system. The meta-rule system has proven extremely valuable while experimenting with a variety of search strategies.

Just as in Fig. 7, each "local transformation" may be regarded as changing the "state" of the Boolean network, or, equivalently, transforming the given Boolean network into an alternate Boolean network. This alternate network realizes the same (given) Boolean logic functions, but at lower cost (cf., Fig. 3). On this view, the objective of the expert system is to find the "state" (i.e., Boolean network) with minimum cost.

The rule-based system thus "searches" the space of all possible states, systematically, under meta-rule control. Suppose the system reaches state s_1 of the state space by applying rule r_1 to node v_1 of the previous Boolean network. The META_RULES subsystem then selects an ordered application sequence which explores a set of "descent estimation boxes." The number of these boxes is a breadth parameter determined by the META_RULES subsystem, as are the breadth and depth parameters which characterize each estimation box. Each such box comprises a subspace of the overall state space. The look-ahead procedure consists of applying a sequence of rules which determines the state of least cost in each descent estimation box. If the box initiated by applying rule r_2 to node v_2 has less estimated cost than that of the boxes associated with the pairs r_3, v_3, r_4, $\cdots$, then state s_2 is selected as the next official state, and the process recurs with s_1 replaced by s_2. In exceptional cases, if the cost of some state, say s_5, is sufficiently low, then the process recurs from s_5 instead of from s_1.

It is significant that the nodes v_2, v_3, v_4 $\cdots$ selected by the META__RULES subsystem are chosen from the immediate neighborhood, in the Boolean network of state s_1, of the original active node v_1. This strategy promotes early competition between rule applications. Such competition is ultimately necessary, since OPTIMIZE terminates only when no further rule applications are possible. Early competition gives the META__RULES subsystem the opportunity to optimize rule tradeoffs.

E. Optimization of Delay

In order to assess the timing benefits of applying a rule, a critical path algorithm is used to compute the typical-case delay times of each signal and the slack time of each gate. Like the critical path finder used in the multilevel synthesis modules, this critical path algorithm is **latent,** i.e., it operates incrementally for greater efficiency. Whenever a transformation is performed, the critical path finder restricts its operation to the gates in the circuit that are descendants or ancestors of an altered gate. This ensures that no time is spent recomputing delay information that didn't change. Since the path finder is working on an actual circuit, the predicted path lengths will now be exact (modulo the accuracy of the library element delay equations) with the exception of the interconnect delays, for which a simple approximation is used.

The user can specify the timing constraints in the form of signal arrival times at the inputs and desired maximum delays at the output. The drive limitations of the logic driving the inputs of the synthesized circuit as well as the loads connected to the output of the synthesized circuit are also specified by the user. The system will synthesize the smallest circuit that meets the timing constraints, or the circuit that comes closest to meeting them. The end product of SOCRATES is a netlist of an automatically generated schematic.

1) Optimization Phases: The process of gate-level optimization proceeds in three main phases. In the first phase, only transformations that reduce both the delay and area cost of the circuit are applied. This phase produces a circuit that is appropriate to the target technology and that strikes a balance between area efficiency and time efficiency. In the second phase, the circuit is optimized for delay only by applying delay-saving rules to the gates along the critical path until the timing constraints are met or the optimizer can do no better. In the last phase, area-saving transformations are used on gates off the critical path. These transformations will reduce the area of the circuit without changing the delay of the circuit.

2) Delay Improving Rules: There are several ways in which local transformations can speed up a circuit.

• *Faster, Similar Gates.* Often logic functions can be implemented in a variety of ways resulting in the same area but different timing behavior. For instance, in CMOS, a NAND/NAND implementation tends to be faster than a NOR/NOR implementation; XNOR gates tend to be faster than XOR gates.

• *Break-Up Large Gates.* Performance improvements can be obtained by replacing larger gates with sequences of smaller, faster gates. The timing benefits of such replacements are strongly a function of the configuration of neighboring gates and can only be assessed by actually recomputing the resulting loading delays of the new circuit.

• *Move Critical Signals Closer to Output.* This is especially useful in cases where one or more input signals arrive late. Reducing the number of levels of logic that such a signal has to traverse to arrive at the output can significantly improve the overall timing of the circuit.

• *Buffering.* Finally, buffering heavily loaded gates appropriately tends to improve the performance of a circuit.

V. Computational Results

We will now describe three types of experimental results. First, the results of applying the SOCRATES system to a set of the ESPRESSO book PLA's [4] which can be thought of as a set of "real world" multiple output, two-level, combinational logic functions. We then demonstrate the flexibility and technology independence of our system with some detailed case studies of specific PLA's (RD53 and F2). Finally, we conclude with some experiments designed to illustrate basic compatibility and detailed interaction of the SYNTHESIS and OPTIMIZE modules.

A. Experiments on ESPRESSO Book PLA's

Both WEAK DIVISION and OPTIMIZE can be run in either area or delay mode. (Note that the other SYNTHESIZE modules, e.g., MULTILEVEL MINIMIZATION and LIBRARY MAPPING, also depend on the cost function and, hence, have both area and delay modes also.) In area mode, the objective is to minimize area independent of delay. However, in delay mode, the objective, roughly speaking, is to minimize area subject to constraints on delay. Table I shows the results for 24 examples. Four experiments were performed on each example:

1) WEAK DIVISION in area mode followed by OP-
   TIMIZE in area mode (A,A);
2) WEAK DIVISION in area mode followed by OP-
   TIMIZE in delay mode (A,D);
3) WEAK DIVISION in delay mode followed by OP-
   TIMIZE in area mode (D,A);
4) WEAK DIVISION in delay mode followed by OP-
   TIMIZE in delay mode (D,D).

We will subsequently refer to the four modes of operation corresponding to these four experiments as AA, AD, DD,

TABLE I
EXAMPLES OF CONSTRAINTS ON DELAY

name(cpu min)	Area (a,a)	(a,d)	(d,a)	(d,d)	Timing (a,a)	(a,d)	(d,a)	(d,d)
mark(0.7)	9	9	21	20	3	3	8	4
f1(1.0)	19	23	20	25	6	5	7	5
clpl(1.0)	20	20	51	48	17	17	12	10
f0(0.4)	26	29	26	29	10	7	10	7
gerf(0.8)	26	25	31	33	12	10	12	10
f2(1.0)	32	32	56	61	6	6	9	4
fadd(1.7)	39	39	46	47	12	12	13	9
adde(1.6)	56	57	56	57	19	18	19	18
dec1(5.4)	73	77	77	84	12	10	10	6
z4(9.3)	76	76	108	117	13	12	22	16
rd53(6.4)	89	95	85	90	22	16	15	11
f5(11.)	97	112	103	124	14	11	14	11
exam(13.)	98	108	112	127	13	10	14	9
f3(8.3)	99	107	99	114	14	13	14	9
f4(9.3)	103	113	108	118	13	10	16	10
8fun(11.)	107	127	125	141	14	13	15	13
plab(12.)	158	176	165	192	18	14	17	14
5xp1(89.)	194	217	220	254	24	19	21	16
dec2(66.)	203	219	215	243	22	17	22	16
f51m(63.)	209	235	256	285	25	19	24	18
root(63.)	234	247	269	287	27	22	43	29
plac(51.)	249	288	303	337	22	17	23	16
bw(192)	286	311	283	307	16	14	17	14
9sym(51.)	361	380	397	426	37	25	50	38
average	119	130	134	148	17	14	18	13

TABLE II
AVERAGES OF AREA AND DELAY VALUES

		WDIV	
	Area/Delay	Area Mode	Speed Mode
OPTIM	Area Mode	119.4/16.6	134.7/18.3
	Speed mode	130.1/13.8	148.6/13.5

TABLE III
EFFECT OF MULTILEVEL MINIMIZATION

name	AA Area	AA Delay	DD Area	DD Delay
fadd	32	9		
adde	59	14		
dec1	69	10		
z4	58	16	61	14
rd53	82	15		
f5	95	14	109	8
exam	94	11	116	10
f4	103	9	124	8
8fun	110	13	128	10
plab	176	15		
dec2	201	20		
plac	256	26	336	15

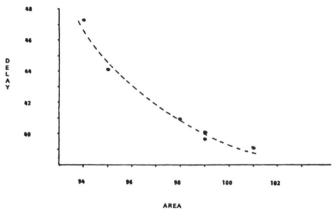

Fig. 10. Area Speed Tradeoffs of RD53.

and DD. In all four experiments, WEAK DIVISION was run with the "unit delay" delay model.

The first half of Table I lists the resulting area values; the second half of the table shows the obtained delay values. Area refers to the number of transistor pairs, delay is in nanoseconds, and (cpu min) refers to the number of VAX 780 cpu minutes to run WEAK DIVISION and OPTIMIZE.

The results are summarized in Table II, which gives the averages of the area and delay values. It is immediately apparent that the best area optimization is obtained by having both modules work in area mode; the best delay optimization is achieved using the two modules in delay mode. Between those two cases, we observe an area increase of 20 percent and a delay decrease of 20 percent. It should be noted that there are some cases (z4, root, 9sym) where a better delay number is obtained for WEAK DIVISION operating in area mode. Section V-C will analyze reasons for these discrepancies and present some partial remedies.

Table III contains a subset of the Table I examples where MULTILEVEL MINIMIZATION was able to further optimize the output. AA corresponds to running MULTILEVEL MINIMIZATION on WEAK DIVISION-A output and then running OPTIMIZE-A. DD corresponds to running MULTILEVEL MINIMIZATION with delay constraints on WEAK DIVISION-D output and then running OPTIMIZE-D.

Comparing the two tables, it can be observed that, in the 12 AA examples, better area numbers (than AA) were obtained in eight cases. Three of the other cases illustrated better area delay tradeoffs. In the 6 DD examples, the delay was reduced (relative to DD) in all but one case (exam2). These numbers indicate the MULTILEVEL MINIMIZATION is often a valuable step to take in the optimization process.

B. Flexibility and Technology Independence

To show the system's flexibility, we took one example and ran it with increasingly severe timing constraints. That is, the required arrival times at circuit outputs were progressively reduced. The results in Fig. 10 show the area/delay tradeoff for example RD53, and demonstrate how SOCRATES can be used to explore a design space quickly and easily.

To demonstrate the technology independence of SOCRATES, when we changed the description of our two input NOR from small and fast to large and slow in the library and ran an example again, SOCRATES replaced all of the original NOR gates with NAND gates and inverters. The schematics for both the original (on the left) and the new circuit (at the right) are shown in Fig. 11. The new implementation without the NOR's is clearly less efficient.

C. SYNTHESIS–OPTIMIZE Interactions

The results of Table I and II clearly establish the overall benefits of the partnership between OPTIMIZE and

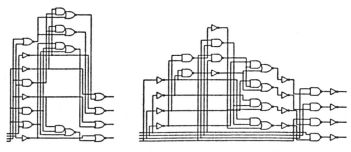

Fig. 11. Example change in technology for F2.

WEAK DIVISION. However, it is of interest to explore further the detailed interaction between modules. One would expect that the AA case (both modules in area mode) would always have the lowest area and that the DD case (both in delay mode) would always have the lowest delay. This is always true in the area case. However, as can be observed in circuits z4, root, and 9sym of Table I, there are discrepancies in the delay case. Examining the differences between AD and DD indicates that when OPTIMIZE is in the delay mode, whichever mode of WEAK DIVISION is used has significant, but mixed, impact on the delay of the final circuit. Subsequent experiments discussed below have shown that, while WEAK DIVISION was always very useful in reducing the area and was effective in reducing the average delay, it was actually detrimental to delay reduction in a few cases.

The delay and area differences between AA and DA or AD and DD can be viewed as indicating how effective WEAK DIVISION is at influencing the delay/area. Similarly, the differences between AA and AD and DA and DD indicate how OPTIMIZE can influence the delay. A tie between AA and AD or DA and DD indicates that the circuit is such that optimizing the area also optimizes the delay. A tie between AD and DD or AA and DA indicates that OPTIMIZE is so effective that it can obtain the best delay/area independent of what WEAK DIVISION does to the input circuit. There are (at least) four possible reasons which may account for these apparent inconsistencies between the WEAK DIVISION and OPTIMIZE components of SOCRATES.

1) limited rule base;
2) WEAK DIVISION unit delay model does not incorporate fan-out or technology dependence;
3) AND/OR translation in the LIBRARY MAPPING module obscures the WEAK DIVISION architecture;
4) different assumptions about forms available for primary inputs

All of these are directly tied to our initial assumption that it would be possible for WEAK DIVISION to operate in a technology-independent manner, using very crude area estimates. Thus, WEAK DIVISION would have the task of efficiently decomposing the original two-level network which could be optimally mapped by OPTIMIZE. Realization that architectural decisions with global implications during WEAK DIVISION were not readily undone

by OPTIMIZE's local transformations inspired the technology-independent unit delay model. Clearly, if there were an infinite rule base, OPTIMIZE could obtain the best final implementation from any representation, but this would require much larger global transformations and may not be feasible in terms of the size of the rule base or in computation time.

As WEAK DIVISION produces a multilevel PLA and OPTIMIZE requires a netlist as input, it is necessary to have a translation program in between them. Based on the above assumptions, the initial LIBRARY MAPPING module used in producing the data of Tables I and II mapped each function in the multilevel cover to a representation consisting of inverters, two input AND gates (AND2), and two input OR gates (OR2). OPTIMIZE views AND2 ad OR2 as dummy gates; they are in the library but have been given very high cost to assure that rules will be applied to replace them. What appears to be happening is that, once larger gates (such as AND4 and XOR) are broken down into two input gates, they are not being recovered by the OPTIMIZE rules so the architecture (or "logic structure") generated by WEAK DIVISION is obscured or lost.

There are (at least) two ways to test and/or remedy this situation. One would be to assure that the rule base contains rules to regenerate gates corresponding to the functions that were decomposed. The second would be to have the input to OPTIMIZE more closely correlate to the output generated by WEAK DIVISION. This requires more technology-dependence in the LIBRARY MAPPING module. After WEAK DIVISION, almost all of the functions satisfy the fan-in constraints of the gates in OPTIMIZE's library. By introducing additional intermediate functions, those that don't satisfy fan-in constraints can be transformed in a straightforward manner to functions that do. A new LIBRARY MAPPING module, Exact__Map, has been written which maps each function into the corresponding gate adding inverters where necessary.

Experimental results with the new LIBRARY MAPPING modules have produced more consistent results. Lower delays have been obtained running all modes, and WEAK DIVISION appears to be having more of an effect on the delay. OPTIMIZE also ran considerably faster with an input consisting of the larger gates. We believe even better area delay tradeoffs could be obtained if there were rules to decompose the large gates in order to investigate alternate configurations.

The original delay model utilized by WEAK DIVISION was the unit delay model. In estimating the delay, functions were decomposed into "gates" satisfying the fan-in constraints of the library and one unit was charged for each gate. (Inverters were not considered.) This model does not take into account load delay, which is often a very important contributor to the total delay through a circuit. Experiments using the library element gate delay model of Fig. 4, which incorporates load delay, have produced more consistent results, in which the WEAK DI-

45

TABLE IV
EFFECT OF INCREASING MODELING DETAIL IN SYNTHESIS MODULE (12 EXAMPLE AVERAGES)

SYNTHESIS Sequence	τ_{CP}(nS)/ Area Ratio				$\Delta\tau$(WD) $\tau_{AD} - \tau_{DD}$	Anomaly Count*
	AA	AD	DA	DD		
WD1/MAP1	13.2/1.9	11.1/1.8	15.5/1.7	11.0/1.5	0.1	6(1)
WD1/MAP2	12.1/1.9	9.6/1.8	11.5/1.8	9.2/1.6	0.4	6
WD1/MAP3	11.3/1.9	9.5/1.8	11.0/1.7	8.8/1.6	0.7	2(3)
WD2/MAP1	12.3/1.9	9.8/1.8	12.9/1.6	9.2/1.5	0.6	4
WD2/MAP3	11.2/1.9	9.6/1.8	10.5/1.5	8.2/1.4	1.4	2
WD3/MAP3	11.2/1.9	9.6/1.8	10.9/1.6	8.7/1.4	0.9	4

$$\text{Area Ratio} = \frac{\text{AREA(phiD)}}{\text{AREA}}.$$

*Anomalies: Individual cases in which WEAK DIVISION pays penalty for area reduction

$$\tau_{AD} - \tau_{DD} < 0.$$

Ties (if any) in parenthesis.

TABLE V
OPTIMIZE WITH AND WITHOUT WEAK DIVISION

Sequence	AA	DD	
- /MAP1/OPT	142.28 11.53	152.07 8.59	(area) $(\tau_{CP}(nS))$
WD2/MAP3/OPT	80.71 11.22	100.21 8.24	
% improvement	43 2.6	34 5.1	

VISION mode has more influence on the final delay. To see how much of an effect load delay has, we recomputed the delay of the Boolean networks created by WEAK DIVISION in unit delay mode. This experiment showed that almost all of them had exceeded the original input delay rather than decreasing it.

Another possible inconsistency between the two programs is that, while WEAK DIVISION assumes that all primary inputs are available in both their positive and complemented forms, OPTIMIZE does not automatically make this assumption. Examining a few examples manually has indicated that a large portion of their delay is due to the complement of a primary input being used in many places, thus having a very high load delay component. By adding additional primary inputs corresponding to the complement of each primary input and specifying they are complements, OPTIMIZE will not need to generate the complement. Early experiments indicate this may reduce the amount of inconsistency.

Tables IV and V show the effect of correcting some of the possible SYNTHESIS module deficiencies discussed above. To this end, we present the results produced by two successively more sophisticated versions of both the WEAK DIVISION and LIBRARY MAPPING modules. We shall refer to the (technology-independent) WEAK DIVISION module, running with the "unit delay" delay model, as "WD1." When the effects of fan-out are introduced as in Fig. 4, but all gates are assumed to have the same set delay, load resistances, and input pin capacitances, we shall mark the experiments "WD2." When the full technology dependence of the "Library Element gate delay" delay model is used, we shall mark the experiments "WD3."

A similar sophistication level is attached to the three versions of the LIBRARY MAPPING module. The Basic AND/OR mapping scheme is marked "MAP1," More sophisticated mapping schemes which try successively harder to retain the logic structure supplied by WEAK DIVISION are marked "MAP2" and "MAP3," respectively.

The results of running a subset of the 24 problems of Tables I and II are presented in Tables IV and V. The rows of Table IV are marked with the appropriate labels for the WEAK DIVISION and LIBRARY MAPPING modules employed. The average delay (in nS) and area ratio constitute the first four columns of data. Area ratio is the OPTIMIZE (delay mode) output area (no WEAK DIVISION, AND/OR mapping) divided by the area obtained using the sequence specified in column 1. This is a measurement of how much WEAK DIVISION influences the final area. This data shows that, generally speaking, final average delays are reduced by increased sophistication of the SYNTHESIS module, while area improvement ratios are slightly reduced. Note that in all cases, the WEAK DIVISION module produces at least a 40 percent area improvement.

The last two columns address the presence of "anomalies," i.e., delay counter-productive WEAK DIVISION. It is observed that increased technology dependence is markedly helpful in reducing the anomaly count, as well as significantly enhancing the delay improvement which can be ascribed solely to the WEAK DIVISION module (1.4 nS i the WD2/MAPS case). Interestingly though, the best results are **not** obtained by the most sophisticated SYNTHESIS sequence.

Finally, Table V summarizes the data comparison with and without WEAK DIVISION. The first row of the table shows the result of **no** WEAK DIVISION. In contrast, the second row gives results for the (best) WD2/MAP3 SYNTHESIS sequence. The third row shows the percent improvement for both area and delay. It can be seen that WEAK DIVISION clearly aids in improving both the area and delay when run in either area or delay mode.

VI. CONCLUSIONS

Both the SYNTHESIS (WEAK DIVISION and MULTILEVEL MINIMIZATION) and OPTIMIZE modules are effective at trading off area for delay. The OPTIMIZE module, which operates with a finite rule set and a heuristic state space search, is effective stand alone, but attains better local minima when aided by the optimized decomposition provided by WEAK DIVISION. It is interesting to note that SYNTHESIS on the average, is effective despite operating with relatively crude delay models. Apparently, this is because SYNTHESIS operates on a higher, more technology-independent level of abstraction. More consistent results, with better delay improvements, can, however, be obtained by adding more technology dependence to the LIBRARY MAPPING modules and incorporating fan-out in the SYNTHESIS delay models (cf., Tables IV and V). These latter results challenge our original paradigm of SYNTHESIS as a purely "architectural" (and, hence, technology-independent) logic optimization process.

In future work, we plan to improve our rule base with more architectural level rules, i.e., ones which are more capable of directly exploiting parallelism to enhance delay. We also plan to investigate still more sophisticated SYNTHESIS modules, including more detailed LIBRARY MAPPING modules, as well as putting WEAK DIVISION in an iteration loop with OPTIMUM PHASE ASSIGNMENT, MULTILEVEL MINIMIZATION, and finally SELECTIVE FLATTENING (which provides WEAK DIVISION with further opportunities for logic optimization).

ACKOWLEDGMENT

We wish to thank D. Gregory, T. Hefner, V. Morgan, and J. Reed for their contributions and helpful comments in regard to the development of the OPTIMIZE rule-based system and the automatic schematics generation. We also wish to acknowledge the technical suggestions and programming assistance of R. Jacoby on the development of WEAK DIVISION and MULTILEVEL MINIMIZATION packages.

REFERENCES

[1] K. A. Bartlett and G. D. Hachtel, "Library specific optimization of multilevel combinational logic," in *Proc. IEEE Int. Conf. on Computer Design*, Oct. 1985.

[2] K. A. Bartlett, W. W. Cohen, A. J. DeGeus, and G. D. Hachtel, "Synthesis and optimization of multilevel logic under timing constraints," in *Proc. IEEE Int. Conf. on CAD*, Nov. 1985, pp. 290–292.

[3] R. K. Brayton and C. T. McMullen, "The decomposition and factorization of Boolean expressions," in *Proc. Int. Symp. on Circuits and Systems*, 1982, pp. 49–54.

[4] R. K. Brayton, G. D. Hachtel, C. T. McMullen, and A. Sangiovanni-Vincentelli, *Logic Minimization Algorithms for VLSI Synthesis*. Hingham, MA: Kluwer Academic Publishers, 1984.

[5] R. K. Brayton, C. L. Chen, C. T. McMullen, R.H.J.M. Otten, and J. Y. Yamour, "Automated implementation of switching functions as dynamic CMOS circuits," in *Custom Integrated Circuits Conf. Proc.*, May 21–23, 1984.

[6] R. K. Brayton and C. T. McMullen, "Synthesis and optimization of multistage logic," in *Proc. IEEE Int. Conf. on Computer Design*, Oct. 1984, pp. 49–54.

[7] R. K. Brayton, private communication, Dec. 1985.

[8] W. W. Cohen, K. A. Bartlett, and A. J. DeGeus, "Impact of metarules in a rule based expert system for gate level optimization," in *Proc. IEEE Int. Symp. on Circuits and Systems*, June 1985, pp. 873–876.

[9] J. Darringer, D. Brand, J. Gerbi, W. Joyner, Jr., and L. Trevillyan, "LSS: A system for production logic synthesis," *IBM J. Res. Dev.*, pp. 537–545, Sept. 1984.

[10] A. J. DeGeus and W. Cohen, "A rule based system for optimizing combinational logic," *IEEE Design and Test of Computers*, pp. 22–32, Aug. 1985.

[11] G. DeMicheli and A. Sangiovanni-Vincentelli, "Multiple constrained folding of programmable logic arrays: Theory and applications," *IEEE Trans. Computer-Aided Design*, vol. CAD-2, pp. 151–167, July 1983.

[12] G. DeMicheli, M. Hoffman, A. R. Newton, and A. Sangiovanni-Vincentelli, "A design system for PLA-based digital circuits," in *Advances in Computer Engineering Design*, A. Sangiovanni-Vincentelli, Ed. Greenwich, CT: Jai Press, 1985.

[13] G. D. Hachtel, A. R. Newton, and A. L. Sangiovanni-Vincentelli, "An algorithm for optimal PLA folding," *IEEE Trans. Computer-Aided Design*, vol. 1, pp. 63–77, Apr. 1982.

[14] M. Hoffman and A. R. Newton, "A domino CMOS logic synthesis system," in *Proc. IEEE Int. Symp. on Circuits and Systems*, June 1985.

[15] R. Rudell, master's report, Dept. EECS, Univ. California, Berkeley, May 1986.

[16] T. Sasao, "An application of multiple-valued logic to a design of masterslice gate arrays," in *Proc. ISMVL-82*, May 1982.

[17] K. Bartlett, R. K. Brayton, G. Hachtel, R. Jacoby, R. Rudell, A. Sangiovanni-Vincentelli, and A. Wang, "Multilevel logic and minimization," in *Proc. IEEE Int. Conf. on CAD*, Oct. 1985.

SYNTHESIS AND OPTIMIZATION OF MULTISTAGE LOGIC

Robert K. Brayton, Curt McMullen

IBM Watson Research Center, Yorktown Hts, NY
Harvard University, Cambridge MA

ABSTRACT

An automatic synthesis system for boolean networks is presented. The system transforms an arbitrary logical description into a set of interconnected circuits implementable in a given target technology. The algorithms are based upon algebraic factorization and boolean minimization. Two notions of boolean division are employed. The procedure has been applied to many practical examples, including a 32-bit microprocessor and a complex ALU.

INTRODUCTION

A complicated logic function is often implemented by multistage circuitry. Although the function is simply a mapping from inputs to outputs, in the actual realization many intermediate signals are created and the computation proceeds in several stages. In the case of arithmetic functions, for example, a logic designer carefully selects intermediates to be created to optimize the performance of his ALU. For control and other data flow logic, the designer's functional description of the unit may not be so closely tied to its most efficient physical implementation. In both cases synthesis tools can be usefully applied to manipulate the logic description.

We describe here an automatic synthesis and minimization procedure for multistage logic, based on algebraic factorization and boolean simplification. The synthesis process is carried out by applying a sequence of operators to a fixed ambient data structure. These operators can both remove intermediates and synthesize new ones, guided by measurements of the frequency of occurence of common subexpressions. If the incoming logic is partly decomposed, the degree to which the initial decomposition is preserved during the synthesis process can be controlled.

Since the synthesis process is automatic and does not require designer intervention, it can be used as a component of a 'silicon compiler'.

The procedure transforms a more or less arbitrary logic description into a set of circuits implementable in a given target technology. It also carries out the optimization of the circuits themselves. In many circuit families, such as Single-Ended Cascode Voltage Switch [1,6,9,10] Differential Cascode [8], and conventional Static CMOS, a fairly complicated logic function (depending on five or more variables) can be implemented as a single circuit. Automatic optimization is essential to the success of these complex circuit families, since design by hand is impractical. For a general discussion of CMOS circuit families, see [5].

Although each function in the final result must be realizable, much of the synthesis procedure is technology independent. Even those operations which are guided by hardware considerations often require only a simple subroutine to check whether or not a given boolean expression is sim-

ple enough to be realized as a single circuit. Due to this modularity, it is easy to change from one circuit family to another. One can also carry out synthesis of a single piece of logic several times, varying the target technology; in this way a comparison between circuit families can be obtained without difficulty.

An outline of the synthesis program is presented below. We begin with a description of boolean networks and the transformations to be performed on them. We then give a more detailed account of the algorithms which carry out these operations, and explain how they fit together to form a synthesis procedure. Finally we demonstrate the effectiveness of this procedure with a practical example.

BOOLEAN NETWORKS

A block of multistage logic (sometimes called a macro) will be referred to as a *boolean network*. It consists of a set of functions, each producing an output of the macro or an intermediate signal, or both. These functions may depend upon the primary inputs to the macro, or upon the intermediates just mentioned, or both. (Of course no cyclic dependency is allowed). A signal which is either a primary input or an intermediate will be referred to simply as a *variable*.

Each function is represented as a sum-of-products boolean expression. Internally this expression is given as a matrix, whose columns correspond to variables and whose rows represent individual product terms.

EXAMPLE.

			A	B	C	D	E
D	$=$	$AB+B\underline{C}$	1	1	2	2	2
			2	1	0	2	2
F	$=$	$D\underline{E}+\underline{D}E$	2	2	2	1	0
			2	2	2	0	1

Here we have used + to denote logical 'or', juxtaposition to denote 'and', and underscore to denote negation. Thus F is defined to be the exclusive or of D and E; note that D is actually an intermediate defined in terms of A and B. In the matrix representation, variables occurring in their positive form are marked with a 1, those occurring in their negative form are marked with a 0, and a 2 indicates that the variable does not occur.

Input to the synthesis system is typically either a symbolic PLA (Programmed Logic Array), a sequence of linked PLAs, or a network of functions produced by a logic description language such as YLL [4]. A single PLA can be interpreted as a boolean network in at least two ways. It can be thought of as a two-stage network, where the product terms on the input plane produce intermediates, sums of which are then used to express the outputs; or as a one-stage network, in which each output is expressed as an

Reprinted from *IEEE Proc. 1984 Int. Conf. on Comp. Des.* (ICCD), pp. 23–28, Oct. 1984.

independent sum of products. In the latter case the creation of intermediate expressions to simplify the network is entirely the responsibility of the synthesis program.

BASIC OPERATIONS

In the course of synthesis a sequence of operators is applied to the boolean network. The operations performed are of four basic types, summarized here and described in detail below.

(1) **Extraction.** A subexpression common to two or more functions is extracted and used to create a new intermediate variable, which then replaces the subexpression wherever it occurs. The new intermediate can be used in both its positive and negative forms.

(2) **Collapsing.** An intermediate of low value is eliminated, and the subexpression it represents is 'pushed back' into the functions which use it. This is the inverse of extraction.

(3) **Simplification.** An unminimized boolean expression is replaced by a logically equivalent but simpler expression. This transformation is essentially 'static'; unlike the other two, it does not usually change the structure of the boolean network.

(4) **Decomposition.** After common subexpressions have been identified by step (1), many of the functions may still be too complex to be implemented in a single circuit of the target technology. Such functions must be further factored and decomposed. This decompostion can be carried out locally without loss of efficiency, because global commonality has already been identified.

After the decomposition step, an actual circuit is automatically designed for each function in the network. This final operation depends on the choice of target technology and will not be described here.

EXTRACTION

The discovery of common subexpressions appearing in the boolean network is the 'creative' step in the automatic synthesis process. It is achieved by forming partial factorizations of each function and generating a list of 'kernels', essential subexpressions from which commonality between two or more functions can always be discovered.

To explain the notion of kernels, we must first introduce the idea of an *algebraic quotient* for a pair of boolean expressions. An *expression* $\iota = f_1 + f_2 + ... + f_n$ is a particular representation of a boolean function as a sum of products. Each single product term f_i is a *cube*. The *algebraic product* of two expressions f and g is only defined when f and g depend upon disjoint sets of variables. It is given simply as the sum of all possible cross-terms $f_i g_j$. Since f and g have disjoint variable sets, no zero products can occur (a variable is never multiplied by its complement). In this sense the product is purely 'algebraic'.

Similarly, the *algebraic quotient* f/g of two expressions is required to depend on variables other than those on which g depends. It is defined to be the largest expression such that

$$f = (f/g)g + r$$

where r is yet another expression (the remainder). Here the product between (f/g) and g is the algebraic product, and the right and left sides of the equation are required to agree as expressions, not just logical functions. That is, the same set of cubes must appear on both sides of the equation.

EXAMPLE.

Let $f = AB + AC + AD + BC + BD$,
$g = A + B$.

Then $f/g = C + D$, since
$f = (A + B)(C + D) + AB$

By using sorting techniques, the computation of the quotient f/g can be carried out very efficiently. In fact the division requires only $O(n \log n)$ steps, where n is the total number of products terms in f and g.

Clearly the formation of quotients is closely related to the problem of factoring an expression. Similarly, the task of identifying subexpressions common to two or more functions is essentially the problem of finding *common divisors*. Here g is a divisor of f if f/g is nontrivial. (Notice that if g is a divisor of f, then so is f/g.) The motivation for defining the *kernels* of an expression is to specify a manageable set of divisors which is still rich enough to allow all common subexpressions to be located.

Definition. The set of *kernels* $\mathscr{K}(f)$ of an expression f is the set of all quotients f/c such that c is a cube and f/c is cube-free.

Here an expression is said to be *cube-free* if no cube can be factored out of it. Thus $ABC + ABD$ is not cube-free (it equals $AB(C+D)$), but $C+D$ alone *is* cube-free.

The number of kernels in an expression is typically much smaller than the number of possible cubes c that can appear in the definition. Indeed, f/c is the empty expression for many choices of c, and for many additional choices it is not cube-free.

Since we represent our boolean expressions in sum-of-products form, it is fairly easy to identify common divisors which are single cubes. It is much harder to locate *cube free* common divisors, and this is the purpose of kernels.

Theorem [3]. If two expressions f and g have a cube-free common divisor, then they have a cube-free divisor which is the sum of cubes common to the expressions h and k, for some $h \in \mathscr{K}(f)$ and $k \in \mathscr{K}(g)$.

Thus to find a subexpression common to two or more functions in a boolean network, we compute the set of all kernels, and then build expressions out of cubes common to pairs of kernels. If no common subexpressions are found in this way, then the above theorem guarantees that the functions in the network are 'relatively prime' -- *they have no cube-free common divisors.* Consequently the only remaining common subexpressions are single cubes; these can be located in a straightforward way.

In practice we limit the calculation to 'depth one' kernels, a somewhat smaller set which still suffices to find useful new intermediates. More details can be found in [3].

Once the candidate intermediates have been identified, they are given ratings, roughly measuring the number of literals they would save in a factored representation of the boolean network. If this rating exceeds a critical value, the intermediate is created and the subexpression it represents is replaced by a single new variable wherever it occurs.

More precisely, to create an intermediate variable v to represent the expression g, we first add the equation $v = g$ to our boolean network. Then we replace each expression f already in the network by the new expression

$$f = v(f/g) + r$$

where r is the remainder resulting from the division of f by g. (Of course if g does not divide f, then f is left unchanged.) We refer to this process as *algebraic substitution* of v for g in f. (Truly boolean substitution will be discussed in the section on simplification below.)

The new intermediate consolidates logic which would otherwise be duplicated in many functions. The rating just mentioned usually reflects a savings in transistors in a physical realization of the network.

We give a simple illustration of the creation of an intermediate.

EXAMPLE.

Consider the three boolean expressions f, g and h given by

$$f = AE+BE+CDE$$
$$g = AD+AE+BD+BE+BF \qquad (1)$$
$$h = A\underline{B}C.$$

We compute the set of kernels for each function:

$$\mathcal{K}(f) = \{ A+B+CD \}$$
$$\mathcal{K}(g) = \{ A+B, D+E, D+E+F \}$$

(The expression h has no kernels). Observing that the expression A+B consists of cubes common to the kernel A+B+CD of f and the kernel A+B of g, we create a new intermediate variable X to represent this expression, and replace A+B with X wherever it occurs. The resulting boolean network is

$$X = A+B$$
$$f = EX+CDE$$
$$g = DX+EX+BF \qquad (2)$$
$$h = C\underline{X}.$$

Notice that the negation of X has been used to simplify the expression for h.

COLLAPSING

Collapsing is the inverse of extraction. If we collapse (push back) the intermediate X in the boolean network (2) above, we obtain the original network (1).

This process is not carried out indiscriminately. Rather, we begin by assigning a value to each intermediate in the boolean network. This value, like the rating used during extraction, measures the degree to which an intermediate helps to simplify the boolean network. We then collapse only those intermediates whose values fall below a certain critical level. This critical level mediates between the otherwise antagonistic extraction and collapsing processes.

Collapsing tends to reduce the total number of circuits and delay stages in the final design. It is also useful in converting a large, sparse network of simple functions into a small, dense network of somewhat larger functions. A sparse network is sometimes produced by the extraction operation; it also might be given as input to the synthesis procedure -- certain logic description languages produce such networks. We will see below that collapsing also enhances the potential for simplification. Thus it is not unusual for collapsing to be invoked several times in the course of synthesis.

SIMPLIFICATION

A single boolean *function* corresponds to many equivalent boolean *expressions*. It is obviously desirable to keep all expressions occuring in a boolean network in as simplified a form as possible. The incoming logical data must be minimized by the synthesis program, and further simplification may be possible after a collapsing operation.

It is slightly less obvious that an expression involving intermediate variables contains implicit relations, which can also be used in its simplification.

EXAMPLE.

$$
\begin{aligned}
Given \quad X &= A\underline{B} + BC\\
Y &= AD,\\
\\
Z &= X + \underline{C}Y \qquad simplifies\ to\\
Z &= X + Y.
\end{aligned}
$$

This simplification results because the condition $C\underline{X}Y$ can never occur; it is part of the *don't-care set* for this network.

In general, to any boolean network we can associate a list of variable combinations which are logically impossible; the totality of all such conditions is the don't-care set for the network. The don't-care set includes, for each intermediate variable v, the condition

$$\overline{v}f + \overline{f}v,$$

where f is the expression represented by v. In the example above, the don't care set is given by the expression

$$\underline{X}(A\underline{B}+BC) + X(\underline{A}\underline{B}+B\underline{C}) + \underline{Y}(AD) + Y(\underline{A}+\underline{D}).$$

These conditions describe implicit relations between the variables in the network, and the exploitation of these relations is essential to the optimization of multistage logic.

The simplification routines we employ were abstracted from the ESPRESSO-II PLA minimizer, described in [2]. These algorithms are based upon the 'unate recursive paradigm', a monotonicity-driven divide and conquer strategy which has proven quite effective. Simplification without a don't-care set is carried out by an fast procedure (called SIMPLIFY in [2]) which does not guarantee that an irredundant cover of prime implicants will be produced, although this is often the case. When minimizing with a don't care set, some time must be invested in the calculation of the offset of the function; once this is done, it is relatively easy to raise implicants to primes and remove redundancies, so these operations are actually carried out.

In practice, the don't-care set for the entire network is too large to process at once, so when minimizing a particular function, we only compute the relations between the variables it directly depends upon. In any case, most of the simplification which can occur is due to this portion of the don't-care set. Note that collapsing tends to bring distant dependencies closer together, so it increases the potential for simplification.

We remark that don't-care simplification can be used to define true *boolean substitution*, in contrast to the algebraic substitution described earlier. We illustrate this notion with a simple example. Suppose we wish to determine if the new intermediate

$$X = \underline{A} + B$$

can be usefully substituted into the existing expression

$$f = AB + \underline{A}C + BC.$$

To do this, we merely simplify f relative to the don't-care set

$$dc = \underline{X}(\underline{A}+B) + X(A\underline{B})$$

associated to the intermediate X.

We begin simplification by determining the minimum subset of the variables A, B, C and X required to express f. Since this procedure is not well known, we describe it in detail. First the onset and offset of f are computed; these are defined as the intersection of f and $\bar{f}$ with the complement of the don't-care set. In this example,

$$on = ABX + \underline{A}CX$$
$$off = \underline{A}\underline{C}X + A\underline{B}\underline{X}.$$

Then the 'blocking matrix' B is computed, as follows:

$$
\begin{array}{c}
\\
B = \\
\\
\\
\end{array}
\begin{array}{cccc}
A & B & C & X \\
1 & 0 & 0 & 0 \\
0 & 1 & 0 & 1 \\
0 & 0 & 1 & 0 \\
1 & 0 & 0 & 1 \\
\end{array}
\begin{array}{cc}
on & off \\
ABX & \underline{A}\underline{C}X \\
ABX & A\underline{B}\underline{X} \\
\underline{A}CX & \underline{A}\underline{C}X \\
\underline{A}CX & A\underline{B}\underline{X} \\
\end{array}
$$

Here B is a 0-1 matrix whose rows correspond to pairs of cubes from the onset and offset of f, and whose columns correspond to variables. Accordingly, each row of B is labelled above by a pair of cubes from the onset and the offset; all possible pairs appear. A '1' appears in the (i,j) position of B if the pair of cubes associated to the ith row *conflict* in the variable associated to jth column. Thus the first row of the matrix reflects the fact that the cubes ABX and $\underline{A}\underline{C}X$ conflict only in the A variable. (If a variable is present in one cube but not the other, no conflict occurs.)

It can be shown that a subset of the variables A, B, C and X is sufficient to express the function f if and only if

the corresponding columns give a 'covering' of the matrix B. (Here a set of columns 'covers' the rows of B if every row has a '1' is some column of that set.) Since we wish to substitute the intermediate X into the expression f, we require that X be included in this subset. One determines by inspection that the minimum column cover containing X corresponds to the variables A, C and X.

We now restrict the onset to this subset of variables, obtaining the new expression $f = AX + \underline{A}CX$. (The superfluous variable B has simply been discarded.) This expression is simplified, using the don't-care set, to obtain the final result

$$f = AX + CX.$$

Note that the algebraic substitution defined earlier could not possibly give rise to this expression, since it involves the term AX even though X depends upon A. It is the result of truly *boolean* substitution.

Boolean substitution (also called strong division) can be a powerful tool for multistage logic synthesis. The following examples of its effectiveness occur in part of a 4-bit ALU description.

$$L = (B\underline{F}+\underline{B}F)(A+E)+\underline{A}\underline{E}(\underline{B}\underline{F}+BF)$$
$$substituted\ into$$
$$R = (B\underline{F}+\underline{B}F)(\underline{A}+\underline{E})+AE(\underline{B}\underline{F}+BF)$$
$$yields$$
$$R = A(E\underline{L}+EL)+\underline{A}(\underline{E}L+E\underline{L})$$

Here both the positive and negative polarities of L have proven useful in the simplification of the expression for R. On the other hand,

$$R = (BF+\underline{B}\underline{F})(\underline{A}+\underline{E})+AE(\underline{B}F+B\underline{F})$$
$$substituted\ into$$
$$L = (BF+\underline{B}\underline{F})(A+E)+\underline{A}\underline{E}(\underline{B}F+B\underline{F})$$
$$yields$$
$$L = A(ER+\underline{E}\underline{R})+\underline{A}(\underline{E}R+E\underline{R})$$

showing that conversely R can be useful in expressing L. Remarkably,

$$T = (DH+\underline{D}\underline{H})((\underline{G}+\underline{C})((\underline{E}+\underline{A})(\underline{B}+\underline{F})+\underline{B}\underline{E})+C G)$$
$$+(D\underline{H}+\underline{D}H)((G+C)(AE(B+F)+BF)+CG)$$
$$substituted\ into$$
$$S=(H+D)((G+C)(AE(B+F)+BF)+CG)+DH$$
$$yields$$
$$S=D(H+\underline{T})+H\underline{T}$$

demonstrating that strong division can sometimes discover unexpected and dramatic simplifications.

Boolean substitution is more powerful that algebraic substitution, but it is also more expensive. In practice, similar results are often obtained more efficiently by algebraic substitution followed by simplification, and this is the technique we actually use. The new intermediate is first incorporated into those existing expressions for which it is an *algebraic* divisor, and then the more expensive boolean simplfication is carried out only for these expressions. In the model example discussed earlier, the algebraic substitution of X into f yields the expression AB+XC. When this is simplified relative to the don't-care set for X, the expression AX+CX found above is again obtained.

DECOMPOSITION

After extraction, it may still be necessary to decompose large functions into smaller pieces. Different technologies demand different decomposition styles; the common problem is to locate within a complicated expression, a large subexpression which can be implemented as a single circuit. Since the task of identifying *global* commonality has already been carried out, this additional decomposition can be carried out for each function independently, without loss of efficiency in the final design.

STEPS IN SYNTHESIS

The operations described above must be linked together in sequence to form a complete synthesis procedure. A typical sequence of operations is as follows:

(1) Collapse incoming data. As usual, those intermediates whose values exceed a given critical amount are not pushed back. By setting this cutoff higher or lower, we can control the degree to which the original decomposition is preserved.

(2) Perform boolean simplification, using the implicit don't care sets.

(3) Extract common subexpressions. Even subexpressions with fairly low value should be extracted, since they help to disclose other subexpressions.

(4) Collapse again. Any intermediates created above which prove to be of little value are now removed.

(5) Simplify again, using don't-care sets.

(6) Decompose the logic into functions simple enough to be implemented in single circuits in the target technology.

(7) Collapse one final time. At this point, do not push back any expression if its removal creates a function which can no longer be realized as a single circuit. At the same time, attempt to reduce the number of circuits and delay stages by trying to push back all but the most valuable intermediates.

(8) Design circuits for each function in the network.

PRACTICAL EXAMPLES

An APL implementation of the synthesis system just described has been in use at Watson Research Center since 1982. An entire IBM 801 32-bit microprocessor and an IBM 370 ALU have been processed with this design tool. Boolean networks with as many as 1000 functions and PLAs with 30 inputs and outputs and over 2000 product terms can be treated with reasonable expenditure of resources. An unminimized boolean network can be synthesized into a circuit family such as Single-Ended Cascode in roughly one to three times the amount of CPU time required to produce a PLA for the same logic. When the input to the synthesis procedure is known to be a minimized PLA, the processing is much faster because many boolean operations can be suppressed.

We illustrate below various stages in the synthesis of a fairly small piece of logic extracted from the IBM System 38 CPU. The incoming data represents six output functions in the form of a single PLA; there are no intermediates initially. The PLA depends upon 27 inputs, which are labelled A though Z, and A'. Intermediates introduced during synthesis will be labelled B', C', etc; the output signals are labelled 1 through 6.

This example illustrates the functioning of extraction, collapsing and a form of decomposition. The original PLA has already been minimized (as a PLA), and consequently simplification has no effect.

The original expressions for the six outputs are shown below. For brevity and comprehensibility, we show below a factored form of each of these expression, although internally each is represented as a sum of products. The factorization algorithm is itself based on kernel generation, so it is not suprising that the factored forms already reveal a good deal of the structure present in the network.

INPUT DATA

$1 = (H\underline{I}+\underline{C}HI)(F(\underline{B}(C\underline{D}+E)+A)+N(B(\underline{C}D+\underline{E})+\underline{A}))$
$\quad + (\underline{E}+\underline{C}D)B(GP(H+I)+\underline{C}KLO)+(E+C\underline{D})\underline{B}(GM(H$
$\quad +I)+\underline{C}J\underline{K}L)+(I+H)G(\underline{A}P+AM)+\underline{C}KL(\underline{A}O+AJ)$

$2 = (H\underline{I}+\underline{C}HI)(F(\underline{S}(\underline{U}(T+V\underline{W})+R)+Q)+N(S(U(\underline{T}$
$\quad +\underline{V}W)+\underline{R})+\underline{Q}))+(I+H)G(M(\underline{S}(\underline{U}(T+V\underline{W})+R)+Q)$
$\quad +P(S(U(\underline{T}+\underline{V}W)+\underline{R})+\underline{Q}))+\underline{C}KL(J(\underline{S}(\underline{U}(T+V\underline{W})$
$\quad +R)+Q)+O(S(U(\underline{T}+\underline{V}W)+\underline{R})+\underline{Q}))$

$3 = (ABCDENQX+\underline{A}\underline{B}\underline{C}\underline{D}EFQX)(\underline{C}HI+\underline{H}\underline{I})+$
$\quad + (ABCDEPQX+\underline{A}\underline{B}\underline{C}\underline{D}EMQX)G(H+I)+$
$\quad + \underline{C}KL(\underline{A}\underline{B}\underline{C}\underline{D}EJQX+ABCDEOQX)$

$4 = ABCDEX(GP(H+I)+\underline{C}(\underline{K}LO+HIN)+\underline{H}\underline{I}N)$

$5 = \underline{A}\underline{B}\underline{C}\underline{D}EX(F(\underline{H}I+\underline{C}HI)+GM(H+I)+\underline{C}J\underline{K}L)$

$6 = (\underline{Z}+\underline{Y})((ABCDENQRSTUVWXA'+$
$\quad + \underline{A}\underline{B}\underline{C}\underline{D}EFQRSTUVWXA')(\underline{C}HI+\underline{H}\underline{I})+$
$\quad + (ABCDEPQRSTUVWXA'+\underline{A}\underline{B}\underline{C}\underline{D}EMQRSTUVWXA')$
$\quad G(H+I)+\underline{C}KL(\underline{A}\underline{B}\underline{C}\underline{D}EJQRSTUVWX\underline{A}'+$
$\quad + ABCDEOQRSTUVWXA'))$

Next we illustrate the results of extraction. As can be seen below, five new intermediate variables have been introduced. In addition, the program has discovered that primary outputs 4 and 5 can also be used to simplify some of the other functions.

RESULT OF EXTRACTION

1	=	$D'(\underline{B}(C\underline{D}+E)+A)+E'(B(\underline{C}D+\underline{E})+\underline{A})$
2	=	$D'(\underline{S}(\underline{U}(T+V\underline{W})+R)+Q)$
		$+E'(S(U(\underline{T}+\underline{V}W)+R)+Q)$
3	=	$QF'+\underline{Q}G'$
6	=	$(QRSTUVWA'F'+\underline{Q}\underline{R}\underline{S}\underline{T}\underline{U}\underline{V}\underline{W}A'G')(\underline{Y}+\underline{Z})$
B'	=	$I+H$
C'	=	$\underline{B}'+\underline{C}HI$
D'	=	$J\underline{H}'+FC'+GMB'$
E'	=	$O\underline{H}'+NC'+GPB'$
4=F'	=	$ABCDEXE'$
5=G'	=	$\underline{A}\underline{B}\underline{C}\underline{D}EXD'$
H'	=	$G+L+K$

The collapsing process reduces the seven intermediates above to four somewhat larger ones, shown below. The degree to which the complexity of the original network has been reduced is self-evident.

RESULT OF COLLAPSING

$$1 = D'(\underline{B}(C\underline{D}+\underline{E})+A)+E'(B(\underline{C}\underline{D}+\underline{E})+\underline{A})$$
$$2 = D'(\underline{S}(\underline{U}(T+V\underline{W})+R)+Q)$$
$$\quad +E'(S(U(\underline{T}+\underline{V}W)+\underline{R})+\underline{Q})$$
$$3 = QF'+\underline{Q}G'$$
$$6 = (QRSTUVWA'F'+\underline{Q}\underline{R}\underline{S}\underline{T}\underline{U}\underline{V}\underline{W}\underline{A}'G')(\underline{Y}+\underline{Z})$$
$$D' = F(\underline{H}\underline{I}+\underline{G}HI)+GM(I+H)+\underline{G}J\underline{K}L$$
$$E' = GP(I+H)+\underline{G}(HIN+\underline{K}\underline{L}O)+\underline{H}\underline{I}N$$
$$4=F' = ABCDEX\underline{E}'$$
$$5=G' = \underline{A}\underline{B}\underline{C}\underline{D}\underline{E}X\underline{D}'$$

This network is now decomposed, with Single-Ended Cascode playing the role of the target technology. In this circuit family the main constraint on feasibility of a given logic expression is the length of its longest product term. The decomposer for this family discovers that if the primary outputs of the network are all complemented, then every expression becomes feasible. (Complementation is allowed in this case because the outputs are feeding a latch). The new network, with the intermediates relabelled, is shown below.

RESULT OF DECOMPOSITION

$$\overline{1} = B'(A(E(C+\underline{D})+\underline{B})+C')+\underline{A}C'(\underline{E}(\underline{C}+D)+B)$$
$$\overline{2} = B'(Q(R(T(V+\underline{W})+\underline{U})+\underline{S})+C')$$
$$\quad +\underline{Q}C'(\underline{R}(\underline{T}(\underline{V}+W)+U)+S)$$
$$\overline{3} = QD'+\underline{Q}E'$$
$$\overline{6} = Q(\underline{R}+\underline{S}+\underline{T}+\underline{U}+\underline{V}+\underline{W}+\underline{A}')+Q(R+S+T+U+V+W+A$$
$$\quad +Q(R+S+T+U+V+W+A')+F'+YZ$$
$$B' = (\underline{H}\underline{I}+\underline{H}\underline{I}+F)\underline{G}(\underline{J}+K+L)+G(\underline{M}(H+I)+\underline{F}\underline{H}\underline{I})$$
$$C' = (\underline{N}+H\underline{I}+\underline{H}I)\underline{G}(K+L+Q)+G(\underline{P}(H+I)+\underline{H}\underline{I}N)$$
$$\overline{4}=D' = \underline{A}+\underline{B}+\underline{C}+\underline{D}+\underline{E}+\underline{X}+C'$$
$$\overline{5}=E' = A+B+C+D+E+X+B'$$

Finally, series-parallel circuits are constructed for each of the functions above. This operation, which is essentially the same as factorization, is also performed by the synthesis system. The series-parallel graph for the intermediate C' is displayed below.

CIRCUIT FOR C'

The boolean network, together with the topological circuit descriptions, is now available for further processing by a layout program such as that described in [7].

REFERENCES

[1] R.K. Brayton, C.L. Chen, C.T. McMullen, R.H.J.M. Otten, Y.J. Yamour, 'Automated Implementation of Switch Functions as Dynamic CMOS Circuits', Proceedings of the 1984 Custom Integrated Circuit Conference, pp. 346-350.

[2] R.K. Brayton, G.D. Hachtel, C.T. McMullen, A. Sangiovanni-Vincentelli, *Logic Minimization Algorithms for VLSI Synthesis*, Kluwer Academic Publishers, 1984.

[3] R.K. Brayton, C.T. McMullen, 'The Decomposition and Factorization of Boolean Expressions', Proceedings of the International Symposium on Circuits and Systems, Rome, 1982, pp. 49-54.

[4] N. Brenner, 'The Yorktown Logic Language: an APL-like Design Language for VLSI Specification and Simulation', these proceedings.

[5] C.L. Chen, R.H.J.M. Otten, 'Considerations for Implementing CMOS Processors', these proceedings.

[6] C.K. Erdelyi, 'Random Logic Design Utilizing Single-Ended Cascode Voltage Switch Circuit in NMOS', Proceedings of the 1984 Custom Integrated Circuit Conference, p. 145.

[7] L.P.P.P. van Ginneken, R.H.J.M. Otten, 'Stepwise Layout Refinement', these proceedings.

[8] L.G. Heller, W.R. Griffin, J.W. Davis, N.G. Thoma, 'Cascode Voltage Switch Logic -A Differential Logic Family', 31st International Solid State Circuits Conference, Digest of Technical Papers, 1984, pp.16-17.

[9] R.H. Krambeck, C.M. Less, H. Law, 'High-Speed Compact Circuits with CMOS', IEEE Journal of Solid-State Circuits, Volume SC-17, No. 3, June 1982, pp. 614-619.

[10] J.A. Pretorius, A.S. Shubat, C.A.T. Salama, D.A. Smith, 'Optimization of Domino CMOS logic and its Applications to Standard Cells', Proceedings of the 1984 Custom Integrated Circuits Conference, p. 150.

S. J. Hong
R. G. Cain
D. L. Ostapko

MINI: **A Heuristic Approach for Logic Minimization**

Abstract: MINI is a heuristic logic minimization technique for many-variable problems. It accepts as input a Boolean logic specification expressed as an input-output table, thus avoiding a long list of minterms. It seeks a minimal implicant solution, without generating all prime implicants, which can be converted to prime implicants if desired. New and effective subprocesses, such as expanding, reshaping, and removing redundancy from cubes, are iterated until there is no further reduction in the solution. The process is general in that it can minimize both conventional logic and logic functions of multi-valued variables.

Introduction

• *Minimization problem*

The classical approach to two-level Boolean logic minimization uses a two-step process which first generates all prime implicants and then obtains a minimal covering. This approach, developed by Quine [1, 2] and McCluskey [3], is a considerable improvement over constructing and comparing all possible solutions. The generation of prime implicants has evolved to a relatively simple process as a result of the efforts of Roth [4], Morreale [5], Slagle et al. [6] and many others. However, the number of prime implicants of one class of n-variable functions is proportional to $3^n/n$ [7]. Thus, for many functions, the number of prime implicants can be very large. In addition, the covering step poses an even greater problem because of its well known computational complexity. Because of the required storage and computations, machine processing to obtain the minimum solution by the classical approach becomes impractical for many-variable problems.

Many attempts have been made to increase the size of problems that can be minimized by sacrificing absolute minimality or modifying the cost function used in covering [6, 8–11]. Su and Dietmeyer [12] and Michalski [13, 14] have reported other serious departures from the classical approach. One recently developed computer program, which essentially represents the state of the art, is said to be able to handle functions of as many as 16 variables [15]. Successful minimization of selected larger functions has also been reported [4, 14]. However, many practical problems of 20 to 30 input variables cannot be handled by the approaches described above and

it does not appear that the classical approach can be easily extended to encompass functions of that size.

• *Heuristic approach*

The approach presented here differs from the classical one in two aspects. First, the cost function is simplified by assigning an equal weight to every implicant. Second, the final solution is obtained from an initial solution by iterative improvement rather than by generating and covering prime implicants.

Limiting the cost function to the number of implicants in the solution has the advantage of eliminating many of the problems associated with local minima. Since only the number of implicants is important, their shapes can be altered as long as the coverage of the minterms remains proper. The methods of modifying the implicants are similar to those that one might use in minimizing a function using a Karnaugh map. The MINI process starts with an initial solution and iteratively improves it. There are three basic modifications that are performed on the implicants of the function. First, each implicant is reduced to the smallest possible size while still maintaining the proper coverage of minterms. Second, the implicants are examined in pairs to see if they can be reshaped by reducing one and enlarging the other by the same set of minterms. Third, each implicant is enlarged to its maximal size and any other implicants that are covered are removed. Thus, both the first process, which may reduce an implicant to nothing, and the third process, which removes covered implicants, may reduce the number of implicants in the solution. The second process facilitates

Reprinted with permission from *IBM Journal of Research and Development,* vol. 18, no. 5, pp. 443–458, Sept. 1974.

the reduction of the solution size that occurs in the other two processes. The order in which the implicants are reduced, reshaped, and enlarged is crucial to the success of the procedure. The details of these processes and the order in which they are applied to the implicants is discussed in later sections. However, the general approach is to iterate through the three main procedures until no further reduction is obtained in the size of the solution.

Our algorithm is designed for minimizing "shallow functions," those functions whose minimal solution contains at most a few hundred implicants regardless of the number of variables. Most practical problems are of this nature because designers usually work with logic specifications that contain no more than a few hundred conditions. The designer is able to express the function as a few hundred implicants because the statement of the problem leads to obvious groupings of minterms. The purpose of the algorithm is to further minimize the representation by considering alternative groupings that may or may not be obvious from the statement of the problem.

To facilitate the manipulation of the implicants in the function, a good representation of the minterms is necessary. The next section describes the cubical notation that is used.

• *Generalized cube format*
The universe of n Boolean variables can be thought of as an n-dimensional space in which each coordinate represents a variable of two values, 0 or 1. A Karnaugh map is an attempt to project this n-dimensional space onto a two-dimensional map, which is usually effective for up to five or six variables. Each lattice point (vertex) in this n-dimensional space represents a *minterm*, and a special collection of these minterms forms an *implicant*, which is seen as a cube of vertices. Following Roth [4], the usual definition of a cube is an n-tuple vector of 0, 1 and X, where 0 means the complement value of the variable, 1 the true value, and X denotes either 0 or 1 or both values of the variable. The following example depicts the meaning of the usual cube notation.

Example 1a
Consider a four-variable (A, B, C and D) universe.

Implicant	Cube notation	Meaning
$\overline{A}\,\overline{B}\,C\,\overline{D}$	0 0 1 0	Minterm with $A = B = D = 0, C = 1$
$A\,\overline{C}$	1 X 0 X	Minterms with $A = 1$, $B = 0$ or 1, $C = 0$, $D = 0$ or 1
U = universe	X X X X	Minterms with $A = 0$ or 1, $B = 0$ or 1, $C = 0$ or 1, $D = 0$ or 1
$\emptyset$ = null	$\emptyset$	No minterms

A more convenient machine representation of 0, 1 and X in the cube is to denote them as binary pairs, i.e., to code 0 as 10, 1 as 01, and X as 11. This representation has the further meaning that 10 is the first of the two values (0) of the variable, 01 is the second value (1), and 11 is the first or the second or both values. Naturally, the code 00 represents no value of the variable and, hence, any cube containing a 00 for any variable position depicts a null cube.

Example 1b
Consider the encoded cube notation of Example 1a.

Cubes	Encoded cubes
0 0 1 0	10 10 01 10
1 X 0 X	01 11 10 11
X X X X	11 11 11 11
$\emptyset$	10 <u>00</u> 11 01

(The 00 entry can be in any variable position. The other values are immaterial.)

We call this encoded cube notation a *positional* cube notation since the positions of the 1's in each binary pair denote the occupied coordinate values of the corresponding variables. With this notation, any non-Boolean variable, which has multiple values, can be accommodated in a straightforward manner. If a variable has t values, the portion corresponding to that variable in the positional cube notation is a binary t-tuple. The positions of each 1 in this t-tuple denote the values of the t-valued variables occupied by the minterms in the cube. Su and Cheung [16] use this positional cube notation for the multiple-value logic. A Boolean variable is a special case of the multiple-value variable.

Consider P variables; let p_i denote the number of values the variable i takes on. We call the p_i-tuple in the positional cube notation the ith *part* of the cube (there are P parts); p_i is called the part size, which is the total number of values there are in the ith coordinate of the P-dimensional multiple-value logic space. Notice that in a cube, the values specified by the 1's in a part are to be ORed, and this constrained part is to be ANDed with other parts to form an implicant.

Any Boolean (binary) output function F with P multiple-value inputs can be mapped into a P-dimensional space by inserting 1's in all points where F must be true and 0's in all points where F must be false. The unspecified points can be filled with d's, meaning the DON'T CARE output conditions. (Often, the 1's and d's are specified and the 0's are filled later.) A list of cubes represents the union of the vertices covered by each cube and is called a cubical cover of the vertices, or simply a *cover*. The goal of the MINI procedure is to cover all of the 1's and none of the 0's with a cover containing a minimum number of cubes. The covers exclusively covering the 1's,

0's, and unspecified points are called, respectively, the ON cover, the OFF cover, and the DON'T CARE cover. When there is no confusion, these covers will be denoted by F, $\bar{F}$, and DC, respectively.

For multiple-Boolean-output functions $(f_1, f_2, \cdots, f_m)$, a tag field [18] has been catenated to the input portion of a cube to denote the multiple-output implicant. We can add an additional m valued dimension for the outputs. This new dimension can be interpreted as representing a multiple-value variable called the *output*. The traditional tag field of an m-tuple binary vector corresponds to our output part in a cube. If the ith bit of the output part is a 1, the ith output is occupied by the cube. We call the whole multiple-output space the generalized universe. Any cube in this universe automatically denotes a multiple-output cube. We denote by F the whole of the multiple-output functions f_1 through f_m. The MINI procedure aims to cover F with a minimal number of cubes in the generalized space.

For generality, we also group input variables into a set multiple-value variables such that the new variables X_i comprising n_i of Boolean input variables have 2^{n_i} values and are called parts. The part sizes are defined as p_i for inputs and m for the output. When groups of inputs are processed through small decoders, the values of decoder output correspond to the multiple values of parts. Each part constitutes a coordinate in the generalized space. The specification of the function is assumed to be a list of regular Boolean cubes with the output tags. The output tag is composed of 0, 1, and d, where 0 means no information, 1 means the cube belongs to the output, and d means the cube is a DON'T CARE for the output. The output side of this specification is the same used by Su and Dietmeyer [12], sometimes known as the output connection matrix.

Example 2a
A Boolean specification and its Karnaugh map.

Inputs				Outputs		
A	B	C	D	f_1	f_2	f_3
0	1	X	X	0	1	0
1	0	X	X	0	1	1
X	0	0	0	1	0	d
X	0	1	1	1	d	1

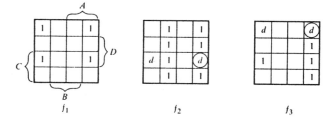

The circled d's in the Karnaugh map show the conflict between 1's and d's. We allow the specification to have conflicts for the sake of enabling the designer to write a concise specification. Any such conflict will be overridden by the d's in our MINI process. Suppose now the inputs are partitioned as $X_1 = \{A, B\}$ and $X_2 = \{C, D\}$. The specification of Example 2a is preprocessed to the generalized positional cube notation as shown below. We call this preprocess a decoding step.

Example 2b
Decoding Boolean specification into the cube format: There are three parts; X_1 and X_2, which take on the four values 00, 01, 10 and 11, and the output, with part size 3. The DON'T CARE cover overrides the ON cover.

X_1	X_2	Output	
0100	1111	010	
0010	1111	011	
1010	1000	100	F
1010	0001	101	
1010	1000	001	
1010	0001	101	DC

The first four cubes for F (ON cover) are the decoded cubes in Example 2a with the output d's replaced with 0's. The last two DC cubes are obtained by decoding only those cubes with d's and replacing the d's with 1's and any non-d output with 0's.

- *Classical concepts in cubical notation*

Several classical concepts have immediate generalizations to the cube structure described in the previous section. The correspondences between a minterm and a point and between an implicant and a cube have already been described. In addition, a prime implicant corresponds to a cube in which no part can admit any more 1's without including some of the $\bar{F}$ space. Such a cube is called a prime cube.

A useful concept in minimization is the size of a cube, which is the number of minterms that the cube contains. It follows from this definition that the size of a cube is independent of the partition of the space into which it is mapped or decoded. Thus, the size of a cube is given by

$$cube\ size = \prod_{i=1}^{P} (\text{number of 1's in part } p_i). \quad (1)$$

Since a cube with one variable per part represents the usual Boolean implicant, each implicant can be mapped into any partitioned cube. Because the resulting cubes can in some cases be merged when the Boolean implicants could not, we have the following theorem.

Theorem 1 The minimum number of cubes that can represent a function in a partitioned space is less than or equal to the minimum number of cubes in the regular Boolean minimization.

To manipulate the cube representation of a function, it is necessary to define the OR, AND, and NOT operations.

1. The OR of two cubes C_1 and C_2 is a list containing C_1 and C_2. The OR of two covers A and B is thus the catenation of the two lists.
2. The AND of two cubes C_1 and C_2 is a cube formed by the bit-by-bit AND of the two cubes. The AND of two covers A and B follows from the above by distributing the AND operation over the OR operation.
3. The NOT of a cube or cover is a list containing the minterms of the universe that are not contained in the cube or cover. The algorithm for constructing this list is discussed in a later section.

The simplest way to decrease the number of cubes of a given problem is to merge some of the cubes in the list. Although this is not a very powerful process, it is well worth applying to the initial specification, especially if there are many entries (a minterm-by-minterm specification is a good example). The following shows the merging of two cubes, which is similar to the merging of two unit-distance Boolean implicants, e.g., $A B \overline{C} \vee A B C = A B$.

Definition The *distance* between two cubes C_1 and C_2 is defined as the number of parts in which C_1 and C_2 differ.

Lemma 1 If C_1 and C_2 are distance one apart, then $C_1 \vee C_2 = C_3$, where C_3 is a bit-by-bit OR of C_1 and C_2.

Proof Let us assume that the difference is in the first part. Let

$$C_1 = a_1 a_2 \cdots a_{p_1} | b_1 b_2 \cdots b_{p_2} | \cdots | n_1 n_2 \cdots n_{p_p}$$

and

$$C_2 = \alpha_1 \alpha_2 \cdots \alpha_{p_1} | b_1 b_2 \cdots b_{p_2} | \cdots | n_1 n_2 \cdots n_{p_p},$$

where $a_i, b_i, \cdots,$ are 0 or 1.   Q.E.D.

The cubes C_1 and C_2 are identical in all but one dimension or part. Therefore, the vertices covered by $C_1 \vee C_2$ can be covered by a single cube with the union of all coordinate values of C_1 and C_2 in that differing coordinate, i.e., $(a_1 \vee \alpha_1), (a_2 \vee \alpha_2), \cdots, (a_{p_1} \vee \alpha_{p_1})$.

The concept of subsumption in cubes is similar to subsumption in the Boolean case ($A B \overline{C} \vee B \overline{C} = B \overline{C}$).

Definition A cube C_2 is said to *cover* another cube C_1 if for every 1 in C_1 there is a corresponding 1 in C_2. In other words, C_1 AND NOT C_2 (bit-by-bit) is all 0's. Since the

cube C_1 is completely contained in C_2, it can be removed from the list, thus reducing the number of cubes of the solution in progress.

Example 3
Consider a three-part example as follows.

$$F = \begin{cases} 1\,0\,1\,0 & 0\,1 & 1\,0 & \text{cube 1} \\ 1\,0\,1\,0 & 1\,0 & 1\,0 & \text{cube 2} \\ 0\,0\,1\,0 & 1\,1 & 1\,0 & \text{cube 3} \end{cases}$$

Cube 1 and cube 2 are distance one apart since they differ only in the second part. The result of merging these two cubes is 1010 11 10, which covers cube 3. Hence, F reduces to one cube, 1010 11 10.

Description of MINI and some theoretical considerations

• *MINI philosophy*
The minimization process starts from the given initial F cover and DC cover (lists of cubes where each cube has P parts). Each part of a cube can be viewed as designating all allowed values of the multiple-valued logic variable, corresponding to that part. The output part can be interpreted as merely another multiple-value variable which may be called the output. When each part's allowed values are ANDed, the resulting cube describes some of the conditions to be satisfied for the given multiple-output logic function corresponding to F and DC specifications. The objective, then, is to minimize the number of cubes for F regardless of the size and shape of the constituent cubes. This corresponds to minimizing only the number of AND gates without fan-in limit, in the regular Boolean two-level AND-OR minimization. We discuss later a simple way of modifying the solution to suite the classical cost criterion.

The basic idea is to merge the cubes in some way toward the minimum number. To do this, MINI first "explodes" the given F cover into a disjoint F cover where the constituent cubes are mutually disjoint. The reasons are

1. To avoid the initial specification dependency. The given cubes may be in an awkward shape to be merged.
2. To introduce a reasonable freedom in clever merging by starting with small, but not prohibitively numerous, fragments such as a minterm list.

The disjoint F is an initial point of the ever decreasing solution. At any point of the process from there on, a guaranteed cover exists as a solution. A subprocess called *disjoint sharp* is used for obtaining the disjoint F.

Given a list of cubes as a solution in progress, a merging of two or more cubes can be accomplished if a larger cube containing these cubes can be found in $F \vee DC$

space to replace them. The more merging is done, the smaller the solution size becomes. We call this the cube *expansion* process. The expansion first orders the given cubes and proceeds down the list until no more merging is possible. This subprocess is not unlike a human circling the "choice" prime implicants in a Karnaugh map. Obviously, one pass through this process is not sufficient.

The next step is to reduce the size of each cube to the smallest possible one. The result of the cube expansion leaves the cubes in near prime sizes. Consequently, some vertices may be covered by many cubes unnecessarily. The cube *reduction* process trims all the cubes in the solution to increase the probability of further merging through another expansion step. Any redundant cube is removed by the reduction and, hence, it also ensures a nonredundant cover.

The trimmed cubes then go through the process called the cube *reshaping*. This process finds all pairs of cubes that can be reshaped into other pairs of disjoint cubes covering the same vertices as before. This step ends the preparation of the solution for another application of cube expansion.

The three subprocesses, expansion, reduction, and reshaping, are iteratively applied until there is no more decrease in the solution size. This is analogous to the trial and error approach used in the Karnaugh map method. We next describe each of these subprocesses and discuss the heuristics used. Brief theoretical considerations are given to formulate new concepts and to justify some of the heuristics used.

• *Disjoint sharp process (complementation)*

The sharp operation $A \# B$, defined as $A \wedge \bar{B}$, is well known. It also yields the complement of A since $\bar{A} = U \# A$, where U denotes the universe. Roth [4] first defined the process to yield the prime implicants of $A\bar{B}$ and used it to generate all prime implicants of F by computing $U \# (U \# (F \vee DC))$. He later reported [17] that Junker modified the process to yield $A\bar{B}$ in mutually disjoint implicants. The operation is easily adapted to our general cubical complex as described in this section.

The disjoint sharp operation. ⊕. is defined as follows: $A \circledast B$ is the same cover as $A\bar{B}$, and the resultant cubes of $A \circledast B$ are mutually disjoint. To obtain this, we give a procedural definition of ⊕ by which $A \circledast B$ can be generated. Consider two cubes $A = \pi_1 \pi_2 \cdots \pi_p$ and $B = \mu_1 \mu_2 \cdots \mu_p$.

Lemma 2 $A \circledast B = C = \vee_{i=1}^{p} C_i$, where C_i is given by

$$C_1 = (\pi_1 \bar{\mu}_1) \pi_2 \pi_3 \cdots \pi_p,$$
$$C_2 = (\pi_1 \mu_1) (\pi_2 \bar{\mu}_2) \pi_3 \cdots \pi_p,$$

$$C_3 = (\pi_1 \mu_1) (\pi_2 \mu_2) (\pi_3 \bar{\mu}_3) \cdots \pi_p,$$
$$\vdots$$
$$C_p = (\pi_1 \mu_1) (\pi_2 \mu_2) (\pi_3 \mu_3) \cdots (\pi_p \bar{\mu}_p), \qquad (2)$$

and AND and NOT operations are performed in a bit-by-bit manner. Whenever any C_i becomes a null cube, i.e., $\pi_i \bar{\mu}_i = \emptyset$, C_i is removed from the $A \circledast B$ list.

Proof It is obvious that the C_i are mutually disjoint; $C = A\bar{B}$ has to be shown. Since for all i, $C_i \subseteq A$ and $C_i \subseteq \bar{B}$, we have $C \subseteq A\bar{B}$. We must show now that every vertex $W \in A\bar{B}$ also belongs to C. Let $W = w_1 w_2 \cdots w_p$; then each w_i is covered by π_i and there exists at least one w_i which is not covered by μ_i. Let the first part where w_i is not covered by μ_i be $\hat{\imath}$. From Eq. (2), we see that $W \in C_{\hat{\imath}}$ and, therefore, $w \in C$. 　　Q.E.D.

Example 4a

A Karnaugh map example for $A = $ universe $= XXXX$ ($\pi_1 \pi_2 \pi_3 \pi_4 = 11\ 11\ 11\ 11$ in our notation) and $B = 11X0$ ($\mu_1 \mu_2 \mu_3 \mu_4 = 01\ 01\ 11\ 10$), the shaded area of the map. Then

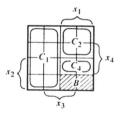

$$C_1 = 0XXX \quad (10\ 11\ 11\ 11),$$
$$C_2 = 10XX \quad (01\ 10\ 11\ 11),$$
$$C_3 = \text{null} \quad (01\ 01\ \underline{00}\ 11) - (\text{delete}),$$
$$C_4 = 11X1 \quad (01\ 01\ 11\ 01).$$

Equation (2) can be expressed more concisely as

$$C_j = (\pi_1 \wedge \mu_1)(\pi_2 \wedge \mu_2) \cdots (\pi_{j-1} \wedge \mu_{j-1})(\pi_j \wedge \bar{\mu}_j)$$
$$\times \pi_{j+1} \pi_{j+2} \cdots \pi_p, \qquad (3)$$

which shows that the μ_j are complemented in order from part 1 through part P. The parts can be complemented in an arbitrary order and still produce a valid $A \circledast B$. Let σ denote an arbitrary permutation on the index set 1 through P. Then Eq. (3) can be rewritten as

$$C_j = (\pi_{\sigma(1)} \wedge \mu_{\sigma(1)})(\pi_{\sigma(2)} \wedge \mu_{\sigma(2)}) \cdots$$
$$(\pi_{\sigma(j-1)} \wedge \mu_{\sigma(j-1)})(\pi_{\sigma(j)} \wedge \bar{\mu}_{\sigma(j)})$$
$$\pi_{\sigma(j+1)} \cdots \pi_{\sigma(p)}. \qquad (4)$$

It is easily shown that the proof of Lemma 2 is still valid if the index set is replaced by the permuted index set. In addition, $A \circledast B$ may be performed for any given permutation and the result will always yield the same number of cubes. However, the shapes of the resultant cubes can vary depending on the part permutation σ, as shown in Example 4b.

Example 4b

Let $A = 1101\ 10\ 11$ and $B = 0101\ 11\ 01$. We calculate A (#) B with two distinct part-permutations, using Eq. (4).

$$A\ (\#)\ B = \begin{Bmatrix} 1000\ 10\ 11 \\ 0101\ 10\ 10 \end{Bmatrix} \text{ if } \sigma = (1, 2, 3)(C_2 = \emptyset);$$

$$A\ (\#)\ B = \begin{Bmatrix} 1101\ 10\ 10 \\ 1000\ 10\ 01 \end{Bmatrix} \text{ if } \sigma = (2, 3, 1)(C_1 = \emptyset).$$

The extension of the (#) operation to include the covers as the left- and the right-side arguments is similar to the regular # case. One difference is that the left-side argument cover F of F (#) G must already be disjoint to produce the desired disjoint $F\overline{G}$ cover. Thus, when $F = \vee f_i$ with the f_i disjoint and g is another cube, F (#) g is defined as

$$F\ (\#)\ g = \vee\ (f_i\ (\#)\ g). \tag{5}$$

If $G = \vee_{j=1}^{n} g_j$, where the g_j are not necessarily disjoint,

$$F\ (\#)\ G = ((\cdots((F\ (\#)\ g_1)\ (\#)\ g_2)\cdots)\ (\#)\ g_{n-1})\ (\#)\ g_n. \tag{6}$$

If F is not in disjoint cubes, the above calculations still produce a cover of $F\overline{G}$, but the resultant cubes may not be disjoint. The proof of the above extensions of (#) is simple and we omit it here.

The definition of F (#) G given in Eq. (6) can be generalized to include the permutation on the cubes of G. One can replace each g_i in Eq. (6) with a permuted indexed $g_{\sigma(i)}$. This cube ordering σ for the right-side argument G influences the shape and the number of resultant cubes in F (#) G. Example 5 illustrates the different outcome of F (#) G depending on the order of the cubes of G.

Example 5

Let F be the universe and G be given as follows.

Produce $\overline{G}$ disjoint $= F$ (#) G.

$$F = 11\ 1111\ 11$$

$$G = \begin{cases} 10\ 1101\ 11 - g_1 \\ 11\ 0010\ 01 - g_2 \end{cases}$$

$$F\ (\#)\ g_1 = \begin{cases} 01\ 1111\ 11 \\ 10\ 0010\ 11 \end{cases}$$

$$(F\ (\#)\ g_1)\ (\#)\ g_2 = \begin{cases} 01\ 1101\ 11 \\ 01\ 0010\ 10 \\ 10\ 0010\ 10 \end{cases}$$

$$F\ (\#)\ g_2 = \begin{cases} 11\ 1101\ 11 \\ 11\ 0010\ 10 \end{cases}$$

$$(F\ (\#)\ g_2)\ (\#)\ g_1 = \begin{cases} 01\ 1101\ 11 \\ 11\ 0010\ 10 \end{cases}$$

The part ordering $\sigma = (1, 2, 3)$ is used for both cases to show the effect of just g_1, g_2 ordering.

As shown by examples 4b and 5, there are two places where permutation of the order of carrying out the (#) process affects the number of cubes in the result. One is the *part* ordering in cube-to-cube (#). and the other is the *right-argument* cube ordering. The choice of these two permutations makes a considerable difference in the number of cubes of F (#) G. Since we obtain $\overline{F}$ as U (#) ($F \vee DC$) and F as U (#) ($\overline{F} \vee DC$) initially, we choose these permutations such that a near minimal number of disjoint cubes will result. The detailed algorithm on how these permutations are selected is presented in a later section. We mention here that these permutations do not affect the outcome in the case of the regular sharp process, because the regular sharp produces *all* prime cubes of the cover. The disjoint $\overline{F}$ obtained in the process of obtaining the disjoint F as above is put through one pass of the cube expansion process (see next section) to quickly reduce the size and thus facilitate the subsequent computations. The $\overline{F}$ used thereafter need not be disjoint.

When the left argument of (#) is the universe, the result is the complement of the right argument. Since we treat the multiple Boolean outputs $f_1, f_2, \cdots, f_m$ as one part of a single generalized function F, we now explain the meaning of $\overline{F}$.

Theorem 2 The output part of $\overline{F}$ represents $\bar{f}_1, \cdots, \bar{f}_m$.

Proof The complementation theorem in [18] states that if $F = \vee E_i f_i$, $\vee E_i = 1$ and $E_i f_i = f_i$, then $\overline{F} = \vee E_i \bar{f}_i$. Let E_i in our case be the whole plane of f_i in the universe; i.e., E_i is a cube denoted by all 1's in every input part and a single 1 in the ith position of the output part. Obviously, $\vee E_i = 1$, $E_i F = E_i f_i = f_i$, and $F = \vee f_i = \vee E_i f_i$. Hence, $\overline{F} = \vee E_i \bar{f}_i = \vee \bar{f}_i$.   Q.E.D.

• *Cube expansion process*

The cube expansion procedure is the crux of the MINI process. It is principally in this step that the number of cubes in the solution decreases. The process examines the cubes one at a time in some order and, from a given cube, finds a prime cube covering it and many of the other cubes in the solution. All the covered cubes are then replaced by this prime cube before a next remaining cube is expanded.

The order of the cubes we process is decided by a simple heuristic algorithm (described later). This ordering tends to put those cubes that are hard to merge with others on the top of the list. Therefore, those cubes that contain any essential vertex are generally put on top of the other cubes. This ordering approximates the idea of taking care of the extremals first in the classical covering step. Thus, a "chew-away-from-the-edges" type of merging pattern evolves from this ordering.

Let S denote the solution in progress; S is a list of cubes which covers all F-care vertices and none of the $\overline{F}$-care vertices, possibly covering some of the DC vertices. Now, from a given cube f of S, we find another cube in $F \lor DC$, if any, that will cover f and hopefully many of the other cubes in S, to replace them. This is accomplished by first expanding the cube f into one prime cube that "looks" the best in a heuristic sense. The local extraction method (see, for instance, [7]), also builds prime cubes around the periphery of a given cube. The purpose there is to find an extremal prime cube in the minimization process. Even though the local extraction approach does not generate all prime cubes of the function, it does generate all prime cubes in the peripheries, which can still be too costly for many-variable problems. To approximate the power of local extraction, the expansion process relies on the cube ordering and other subprocesses to follow. Since only one prime cube is grown and no branching is necessary, the cube expansion process requires considerably less computation than the local extraction process.

The expansion of a cube is done one part at a time. We denote by $SPE(f; k)$ the single-part expansion of f along part k; SPE can be viewed as a generalized implementation of Roth's coface operation on variable k.

Definition Two disjoint cubes A and B are called k-*conjugates* if and only if A and B have only one part k where the intersection is null; i.e., when part k of both A and B is replaced with all 1's, the resultant cubes are no longer disjoint.

Example 6

Let f be $101X$ in regular Boolean cube notation. The cubes $0X11$, $X1XX$ and 1000 are examples of 1-, 2- and 3-conjugates of f, respectively. There is no 4-conjugate of f in this case.

Let $H(f; k)$ be the set of all cubes in $\bar{F}$ that are k-conjugates of the given cube f in S.

$$H = \{g_i | f \text{ and } g_i \text{ are } k\text{-conjugates}\}, \qquad (7)$$

where we assume that the $\bar{F}$ is available as $\bar{F} = \lor g_i$, which is obtained as a by-product of the disjoint F calculation. (Since the cube expansion process makes use of $\bar{F}$, we say that S is expanded against $\bar{F}$.) Further denote as $Z(f; k)$ the bit-by-bit OR of the part k of all cubes in $H(f; k)$. When $H(f; k)$ is a null set, $Z(f; k)$ is all 0's. The single-part expansion of $f = \pi_1 \pi_2 \cdots \pi_k \cdots \pi_p$ along part k is defined as

$$SPE(f; k) = \pi_1 \pi_2 \cdots \pi_{k-1} \overline{Z(f; k)} \; \pi_{k+1} \cdots \pi_p, \qquad (8)$$

where $\bar{Z}$ denotes bit-by-bit complementation.

Example 7a

Let f and $\bar{F}$ be as follows. The SPE along parts 1, 2 and 3 is obtained.

$$f = 10\ 10\ 0110$$
$$\bar{F} = \begin{cases} 11\ 11\ 1000 = g_1, \\ 11\ 01\ 0011 = g_2, \\ 01\ 10\ 0001 = g_3. \end{cases}$$

Then,

$$H(f; 1) = \emptyset \text{ and } Z(f; 1) = 00$$

$$SPE(f; 1) = \underline{11}\ 10\ 0110,$$

$$H(f; 2) = \{g_2\} \text{ and } Z(f; 2) = 01$$

$$SPE(f; 2) = 10\ \underline{10}\ 0110 = f,$$

$$H(f; 3) = \{g_1\} \text{ and } Z(f; 3) = 1000$$

$$SPE(f; 3) = 10\ 10\ \underline{0111}.$$

Example 7b

In the regular Boolean case, let $f = 101X$ and let $\bar{F} = g_1 \lor g_2 \lor g_3 = \{00X0, 0XX1, 110X\}$ as shown in the Karnaugh map below.

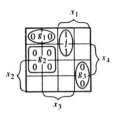

K	$H(f; k)$	$\overline{Z(f; k)}$	$SPE(f; k)$
1	g_1, g_2	1	$\underline{1}01X = f$
2	null	X	$1\underline{X}1X$
3	null	X	$10\underline{X}X$
4	null	X	$101\underline{X} = f$

Notice that the Boolean case is a degenerate case where 1 or 0 in any variable can stay the same or become an X when the coface operation succeeds.

Lemma 3 Let C be any cube in $F \lor DC$ which contains the cube f of S. Then part k of C is covered by part k of $SPE(f; k)$. That is, if $C = \mu_1 \mu_2 \cdots \mu_k \cdots \mu_p$, the 1's in μ_k are a subset of the 1's in $\overline{Z(f; k)}$.

Proof Suppose μ_k contains a 1 that is not in $\overline{Z(f; k)}$. This implies $\mu_k \cdot Z(f; k) \neq \emptyset$, which in turn implies that there exists a cube g in $\bar{F}$ which is a k-conjugate of f and part k of g has a non-null intersection with μ_k. Since C covers f, and f and g have non-null intersection in every part but k, C and g have non-null intersection. This contradicts the hypothesis that C is in $F \lor DC$.   Q.E.D.

It follows from the above that part k of $SPE(f; k)$ is prime in the sense that no other cube in $F \lor DC$ containing f can have any more 1's in part k than $SPE(f; k)$ has. We define part k, $\overline{Z(f; k)}$, or $SPE(f; k)$ as a *prime* part, which leads to the following observation.

Theorem 3 A cube is prime if and only if every part of the cube is a prime part.

A cube can be expanded in every part by repeatedly applying the *SPE* as follows:

$expand(f) =$

$$SPE(\cdots SPE(SPE(SPE(f; 1); 2); 3) \cdots; p). \quad (9)$$

To be more general, let σ be an arbitrary permutation on the index set 1 through p; then

$$expand(f) = SPE(\cdots SPE(SPE(SPE(f; \sigma(1)); \sigma(2));$$

$$\sigma(3))\cdots; \sigma(p)). \quad (10)$$

The result of $expand(f)$ may not be distinct for distinct part permutations. However, the part permutation does influence the shape of the expanded cube, and each expansion defines a prime cube containing f by Theorem 3.

Example 8
Let $f, \bar{F}$ and the part permutations be as follows.

$f = $ 01 1000 10 | *part permutation* : $expand(f)$

$$\bar{F} = \begin{cases} 11\ 0100\ 11 \mid \sigma_1 = (1, 2, 3) & : 11\ 1011\ 10 : A \\ 10\ 0011\ 01 \mid \sigma_2 = (1, 3, 2) & : 11\ 1000\ 11 : B \\ 01\ 0110\ 01 \mid \sigma_3 = (3, 2, 1) & : 01\ 1001\ 11 : C \end{cases}$$

The part permutations $(2, 1, 3)$ and $(2, 3, 1)$ both produce A and $(3, 1, 2)$ produces B.

There is no guarantee of generating all prime cubes containing f even if all possible part permutations are used unless, of course, f happens to belong to an essential prime cube. The goal is not to generate prime cubes but rather to generate an efficient cover of the function. Therefore, a heuristic procedure is used to choose a permutation for which $expand(f)$ covers as many cubes of S as possible. Consider a cube $C(f)$ defined by

$$C(f) = \overline{Z(f; 1)}\ \overline{Z(f; 2)} \cdots \overline{Z(f; P)}. \quad (11)$$

For Example 8, $C(f)$ is 11 1011 11. Obviously, $C(f)$ is not always contained in $F \vee DC$. However, any expansion of f can at best eliminate those cubes of S that are covered by $C(f)$ which is called the *over-expanded* cube of f. The permutation we choose is derived from examining the set of cubes of S that are covered by $C(f)$.

Let the *super* cube C of a set of cubes $T = \{C_i | i \in I\}$ be the smallest cube which contains all of the C_i of T. We state the following lemma omitting the proof.

Lemma 4 The super cube C of T is the bit-by-bit OR of all the C_i of T.

One can readily observe that $C(f)$ is a super cube of all prime cubes that cover f and is also the super cube of the set of cubes $\{SPE(f; k) | k = 1, 2, \cdots, P\}$.

For a given $f \in S$, let $f' = expand(f)$ obtained with a chosen part permutation. If f' covers a subset of cubes S' of S, $f \in S'$, the whole set S' can be replaced by f', which decreases the solution size. If, instead of f', one uses a super cube f'' of S' in the replacement, the reduction of the solution is not affected. The reason for using f'' is that $f'' \subseteq f'$, which implies that f'' has a higher probability of being contained in another expanded cube of S than f' does. Of course, f'' may not be a prime cube. In the next section we show how this f'' is further reduced to the smallest necessary size cube that can replace the S'.

The cube expansion process terminates when all remaining cubes of S are expanded. The expansion process described above also provides an alternate definition of an essential prime cube.

Theorem 4 The cube $expand(f)$ of a vertex f is an essential prime cube (EPC) if and only if $expand(f)$ equals the over-expanded super cube $C(f)$. It follows that when $expand(f)$ is an essential prime cube, the order of part expansion is immaterial. (Proof follows from the remark after Lemma 4.)

• *Cube reduction process*
The smaller the size of a cube, the more likely that it will be covered by another expanded cube. The expansion process leaves the solution in near-prime cubes. Therefore, it is important to examine ways of reducing the size of cubes in S without affecting the coverage. Define the *essential* vertices of a cube as those vertices that are in F and are not covered by any other cube in S. Let f' be the supercube of all the essential vertices of a cube $f \in S$; then f' is the smallest cube contained in f which can replace f in S without affecting the solution size. Of course, if f does not contain any essential vertices, then the reduced cube is a null cube and f may be removed from the S list, decreasing the solution size by one. Let $S = f \vee \{S_i | i \in I\}$; then the reduced cube f' can be obtained as

$$f' = \text{the super cube of } f \ \textcircled{\#} \ ((\bigvee_{i \in I} S_i) \vee DC). \quad (12)$$

In Eq. (12) a regular # operation can be used in place of $\textcircled{\#}$. In fact, the irredundant cover method [5, 7, 12] uses the regular # operation between a given cube and the rest of the cubes of a solution. The purpose of this # operation in the irredundant cover method is to remove a redundant cube. In our case the reduction of the size of the given cube is the primary purpose. Regardless of the purpose, we claim that the use of $\textcircled{\#}$ facilitates this type of computation in general. The number of disjoint cubes of a cover is usually much smaller than the number of all prime cubes of the same cover, which is the product of regular # operations.

In our programs the reduced cube f' is not obtained in the manner suggested by Eq. (12). Since the super cube is the desired result, a simpler tree type algorithm can be used to determine the appropriate reduction of each part of the given f.

The cube reduction process goes through the list of cubes in the solution S in a selected order and reduces each of them. The cube ordering algorithm for the reduction step is a heuristic way to maximize the total cube size reductions; the process removes the redundant cubes and trims the remaining ones.

In the previous section, we mentioned how the replacement cube (f''') was found by the expansion process. The size of this cube can be further reduced along with the sizes of the remaining cubes in the solution. We do this within the cube expansion process by first reducing all the remaining cubes in the solution one at a time against the replacement cube f''', and then reducing the f''' to the smallest necessary size. This is illustrated in the following example.

Example 9

Let the replacement cube f''' (the shaded area) and some of the remaining cubes of S in the periphery of f''' be as shown on the left below. The right side shows the desired cube shapes before the expansion process proceeds to the next cube in the solution.

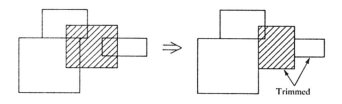

The reduction of one cube A against another cube B assumes that B does not cover A and that the two differ in at least two parts. Cube A can be reduced if and only if all parts of B cover A except in one part; let that be part j. Given $A = \pi_1\pi_2\cdots\pi_j\cdots\pi_p$ and $B = \mu_1\mu_2\cdots\mu_j\cdots\mu_p$, the trimmed A becomes $A' = \pi_1\pi_2\cdots(\pi_j\bar{\mu}_j)\cdots\pi_p$.

• *Cube reshaping process*

After the expansion and reduction steps are performed, the solution in progress contains minimal vertex sharing cubes. The nature of the cubes in S is that there is no cube in $F \lor DC$ that covers more than one cube of S. Now we attempt to change the shapes of the cubes without changing their coverage or number. What we adopted is a very limited way of reorganizing the cube shapes, called the cube *reshaping* process. Considering that the reshapable cubes must be severely constrained, it was our surprise to see significant reshaping taking place in the course of minimization runs on large, practical functions.

The reshaping transforms a pair of cubes into another disjoint pair such that the vertex coverage is not affected. Let us assume that S is the solution in progress in which no cube covers another and the distance between any two cubes is greater than or equal to two. Let A and B be

two cubes in S. Then the cubes $A = \pi_1\pi_2\cdots\pi_p$ and $B = \mu_1\mu_2\cdots\mu_p$ in that order are said to satisfy the *reshaping condition* if and only if

1. The distance between A and B is exactly two.
2. One part of A covers the corresponding part of B.

Let i and j be the two parts in which A and B differ and let j be the part in which A covers B, i.e., π_j covers μ_j: π_i cannot cover μ_i for, if it did, then A would cover B. The two cubes

$$A' = \pi_1\pi_2\cdots\pi_i\cdots(\pi_j\land\bar{\mu}_j)\cdots\pi_p \qquad (13)$$

and

$$B' = \pi_1\pi_2\cdots(\pi_i\lor\mu_i)\cdots\mu_j\cdots\mu_p \qquad (14)$$

are called the reshaped cubes of A and B. The process is called *reshape* $(A; B)$.

Lemma 5 The reshaped cubes A' and B' are disjoint and $A \lor B = A' \lor B'$.

Proof The jth part of A' is $\pi_j\land\bar{\mu}_j$ (bit-by-bit) and the jth part of B' is μ_j; hence, A' and B' are disjoint. In reshaping, A is split into two cubes A' and $A'' = \pi_1\pi_2\cdots(\pi_j\land\mu_j)\cdots\pi_p$. But $(\pi_j\land\mu_j) = \mu_j$ because π_j covers μ_j; thus the distance between A'' and B is one. So A'' and B merge into the single cube B'.　　Q.E.D.

The reshape operation between A and B is order dependent. If the cubes in S are not trimmed, it may be possible to perform reshape in either of two ways (e.g., if A and B are distance two apart, $\pi_i \supset \mu_i$ and $\pi_j \subset \mu_j$). Since A is split and one part is merged with another cube B, the natural order would be the larger cube first and the smaller cube second when checking the conditions for reshaping. After reshaping, A', the remaining part of A, has a greater probability of merging since it has been reduced in size.

Example 10a

Let S consist of three cubes A, B and C as follows.

$$S = \begin{cases} 11\ 0110\ 01 : A \\ 10\ 1001\ 01 : B \\ 01\ 0110\ 10 : C \end{cases}$$

A and B satisfy the reshaping condition to yield

$A' = 01\ 0110\ 01$ (can merge with C in the next expansion step)

$B' = 10\ 1111\ 01$

Or A and C may yield

$A' = 10\ 0110\ 01$ (can merge with B)

$C' = 01\ 0110\ 11$

Example 10b

Regular Boolean Karnaugh map example.

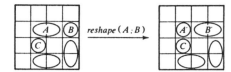

A' can merge with C in the next expansion step. Or A and C could be reshaped to A' and C' such that A' and B can merge later.

The reshaping operation can be viewed as a special case of the consensus operation. Notice that the reshaped cube B' is the consensus term between A and B. The reshaping condition holds only if the pair of cubes can be represented by a consensus cube plus another cube for the remainder of the vertices covered by A and B. The consensus operation is used in classical minimization methods to generate prime implicants from a given implicant list of functions.

Algorithmic description of MINI

This section describes the algorithms which implement the procedures outlined in the previous section. The algorithms are intended as a level of description of MINI which is between the theoretical considerations and a real program. They show the flow of various subprocesses and the management of many heuristics.

• *Main procedure*

M1. Accept the Boolean specification.

M2. Accept the partition description.

M3. Extract the F-care specification and decode into cubes according to the partition description. Assign to F.

M4. Generate DON'T CARE specification (DC) due to any inputs which appear in more than one part.

M5. Extract the original DON'T CARE specification and add to the DC specification generated in M4.

M6. Decode the DC specification into cubes. Assign to DC.

M7. Generate the partition description in the format required by subsequent programs.

M8. Let F be the distance one merging of $F \vee DC$.

M9. Let $\bar{F}$ be $U \ⓖ\ F$.

M10. Let $\bar{F}$ be $\bar{F}$ expanded against F.

M11. Generate the disjoint F by $U \ⓖ\ (\bar{F} \vee DC)$.

M12. Let F be F expanded against $\bar{F}$ and compute the solution size.

M13. Reduce each cube of F against the other cubes in $F \vee DC$.

M14. Reshape F.

M15. Let F be F expanded against $\bar{F}$ and compute the solution size.

M16. If the size of the new solution is smaller than the size of the solution immediately before the last execution of M13, go to M13. Otherwise F is the solution.

Remarks

M3 and M6 give the F and the DC covers, respectively. M8 is performed for computational advantage only. M9 produces the disjoint $\bar{F}$ cover. M10 is for computational advantage. M11 produces the disjoint F cover. M13 − 16 form the main loop which produces decreasing size solutions which contain all of the F vertices and perhaps some of the DC vertices. M1 through M6 are for the Boolean specified functions. If the original specification is in cube notation for F and DC, the procedure should start at M7.

• *Preparatory algorithms*

Assume that the function is given as a Boolean specification and that the partition information is given as a permuted input list and a part size list. The sum of the numbers in the input part size list should equal the length of the permuted input list. For example, the input variable permutation (0 1 3 5 4 2 1) and the decoder sizes (2 2 3) imply that the inputs are partitioned as (0, 1), (3, 5), and (4, 2, 1). The variable number 1 is assigned to both the first and the third parts. The order of variables within a part does not influence the minimization, but it does influence the bit pattern in the part.

Separate F and DC specifications.

F specification:

PF1. Eliminate from the original specification all rows that do not contain a 1 in the output portion.

PF2. Replace each d (DON'T CARE symbol) in the output portion with a 0.

DC specification:

PDC1. Select all rows of the original specification that contain at least one d in the output portion.

PDC2. In the output portion, replace each 1 with a 0.

PDC3. In the output portion, replace each d with a 1.

Decode the given Boolean specification of F (the output portion now contains only 1's and 0's) into a cubical representation with the given partition.

PD1. Construct a matrix G whose kth column is the column of the input specification that corresponds to the kth variable in the permuted input list.

PD2. Perform steps PD2 − 6 for each input part from first to last.

PD3. Let NF be the first p columns of G, where p is the number of variables in the part.

PD4. Drop the first p columns of G.

PD5. Decode each row of NF into the bit string of length 2^p, which corresponds to the truth table entries for that row. (For example, −10 becomes 001000010.)

PD6. For the next part, go to PD3.

PD7. The output portion of the specification without change becomes the output part of the decoded cubes.

Generate, for M4, the DC specification due to the assignment of an input to more than one part. The result has the form of the matrix G given in PD1. The output part of each of the resultant cubes contains all 1's.

PMDC1. Start with a null *DC*.

PMDC2. Repeat steps PMDC3–7 for all variables.

PMDC3. If variable *I* appears in *k* parts, generate the following $2^k - 2$ rows of *DC*. The matrix *M* gives the input parts and the output parts are all 1's.

PMDC4. Let *M* be a matrix of −'s with dimensions $(2^k - 2)$ rows × (length of the permuted input list) columns.

PMDC5. Let *W* be a $(2^k - 2) \times k$ matrix of 1's and 0's where the *n*th row represents the binary value of *n*. The all-0 and all-1 rows are not present.

PMDC6. The *t*th column of *W* replaces the column of *M* that corresponds to the *t*th occurrence of variable *I* in the permuted input list.

PMDC7. For the next variable, go to PMDC3.

• *Distance one merging of cubes*

S1. Consider the cubes as binary numbers and reorder them in ascending (or descending) order by their binary values.

S2. The bits in part *k* (initially *k* = *P*, the last part) are the least significant ones. Starting from the top cube, compare adjacent cubes. If they are the same except in part *k*, remove both cubes and replace them with the bit-by-bit OR of the two cubes. Proceed through the entire list.

S3. Reorder the cubes using only the bits in part *k*. In case of a tie, preserve the previous order.

S4. Part *k* − 1 now contains the least significant bits. Let *k* be *k* − 1 and go to S2. Terminate when all parts have been processed.

Remark

If any set of cubes are distance one apart and the difference is in part *k*, the set of cubes will appear in a cluster when ordered using the bits of part *k* as the least significant positions.

• *Disjoint sharp of a cover F against a cover G* ($F \circledast G$)

Ordering of right side argument; reorder cubes of *G*:

ORDG1. Sum the number of 1's in each part of the list *G* and divide by the part size to obtain the average density of 1's in each part.

ORDG2. For each part, starting from the most dense to the least dense, do steps ORDG3–6.

ORDG3. Sum the number of 1's per bit position for every bit in the part. Order the bits from most 1's to least 1's.

ORDG4. Do steps ORDG5, 6 for all bits in the part in the order computed in ORDG3.

ORDG5. Reorder the cubes of *G* such that the cubes with a 1 in the bit position appear on top of the cubes with a 0 in the bit position. Within the two sets, preserve the previous order.

ORDG6. Go to ORDG5 for the next bit of the part. If all bits in a part are done, go to ORDG3 for the next part.

ORDG7. Terminate when the last bit of the last part has been processed.

Remarks

The ordering procedure has been obtained from numerous experiments. The objective is to order *G* such that the number of cubes produced by the disjoint sharp will be as small as possible. One of the properties of the above ordering is that it tends to put the larger cubes on top of the smaller cubes.

$F \circledast G$:

DSH1. Order *G* according to ORDG.

DSH2. Remove the first cube of *G* and assign it to the current cube (*CW*).

DSH3. Let *Z* be the list of cubes in *F* which are disjoint from *CW*. Remove *Z* from *F*.

DSH4. Compute the internal part ordering for $F \circledast CW$ as follows: For each part compute the number of cubes in *G* that are disjoint from *CW* in that part. Order the parts such that the number of cubes that are disjoint in that part are in descending order.

DSH5. Using Eq. (4), compute $F \circledast CW$ with the part permutation given by DSH4; then add the result to the *Z* list.

DSH6. If *G* is empty, the process terminates and the *Z* list is the result. If *G* is not empty, let *F* be the *Z* list and go to DSH2.

Remarks

ORDG and DSH4 are the two heuristic ordering schemes used in the sharp process. These two heuristics were chosen so that the disjoint sharp process would produce a small number of disjoint cubes.

• *Expansion of F against G*

Ordering the cubes of *F*:

ORDF1. Sum the number of 1's in every bit position of *F*.

ORDF2. For every cube in *F*, obtain the weight of the cube as the inner product of the cube and the column sums.

ORDF3. Order the cubes in F such that their weights are in ascending order.

Remarks

This ordering tends to place on top of the list those cubes that are hard to merge with other cubes. If a cube can expand to cover many other cubes, the cube must have 1's where many other cubes have 1's, and hence its weight is large. This heuristic ordering produces the effect of "chewing-away-from-the-edges." When there is a huge DON'T CARE space, $F \vee DC$ can be used instead of F in ORDF1, for more effective expansion of cubes.

The expansion process:

EXP1. Order the cubes of F according to ORDF.

EXP2. Process the unexpanded cubes of F in order. Let f be the current cube to be expanded.

EXP3. For each part k, compute the k-conjugate sets $H(f;k)$ given by Eq. (5) and their $Z(f;k)$; then form the over-expanded cube $C(f)$ given by Eq. (9).

EXP4. Let Y be the set of cubes of F that are covered by $C(f)$.

EXP5. For each part, compute the weight as the number of cubes in Y whose part k is covered by part k of f.

EXP6. Order the parts in ascending order of their weights.

EXP7. Let ZW be the expanded f using the above part permutation and Eq. (8).

EXP8. Let Y be all of the cubes of F that are covered by ZW and remove Y from F.

EXP9. Let S be the super cube of Y.

EXP10. Find all cubes in F that are covered by ZW in all parts but one. Let these cubes by Y.

EXP11. Reduce each cube of Y against ZW.

EXP12. Let T be the super cube of Y. Let ZW be $ZW \wedge (S \vee T)$.

EXP13. The modified expanded f is ZW. Append ZW to the bottom of F.

EXP14. If there are any unexpanded cubes in F, go to EXP2.

Remarks

EXP3–6 defines the internal part permutation. The idea is to expand first those parts that, when expanded, will cover the most cubes that were not covered by the original cube. EXP8 removes all covered cubes. The S of EXP9, which is contained in ZW, could replace f now if a cube reduction were not employed. By EXP10–11, all remaining cubes of F are reduced. The intersection of T and ZW denotes the bits of the initial expanded prime cube ZW which were necessary in the reduction of any cube. The final replacement for the original cube F is thus $(ZW \wedge T) \vee S = ZW \wedge (T \vee S)$. The cube that re-

places f is the smallest subcube of a prime cube containing f that can contain and reduce the same cubes of F that the prime cube can.

• *Reduction of cubes*

The actual experimental program for this algorithm is quite different from a straightforward disjoint sharp process. For efficient computation a tree method of determining essential bits of a cube is used. The algorithm given below is only a conceptual one. First the cubes to be reduced (given as F) are reordered according to ORDF except that ORDF3 is modified to order cubes in descending order of their weights. This ordering tends to put cubes that have many bits in common with other cubes on top of the list. It is assumed that the DC list is also given.

RED1. Order the cubes of F with the modified ORDF.

RED2. Do steps RED2–4 for all cubes of F in order. Let the current cube be f.

RED3. Replace f with the super cube of the disjoint sharp of f against $DC \vee (F - f)$; $F - f$ denotes all the cubes of F except f. If the super cube is a null cube, f is simply removed from the list.

RED4. Go to RED2 for the next cube.

• *Reshape the cubes of F*

RESH1. Order the cubes of F by the modified ORDG used in RED1.

RESH2. Do for all cubes of F in order. Let the current cube be $C1$.

RESH3. Proceed through the cubes below $C1$ one at a time until a reshape occurs or until the last cube is processed. Let the current cube be $C2$.

RESH4. If $C1$ covers $C2$, remove $C2$ from F and go to RESH3.

RESH5. If $C1$ and $C2$ are distance one apart, remove $C1$ and replace $C2$ with $C1$ bit-by-bit OR $C2$ and mark the ORed entry as reshaped. Go to RESH2.

RESH6. If $C1$ and $C2$ do not meet the reshaping condition, go to RESH3.

RESH7. If $C1$ and $C2$ meet the reshaping condition, form the reshaped cubes $C1'$ and $C2'$. Replace $C2$ with $C1'$ and $C1$ with $C2'$. Mark these cubes as reshaped. Go to RESH2.

RESH8. If $C1$ is not the last cube in the list, go to RESH2.

RESH9. Let all reshaped cubes be R and all unchanged cubes be T.

RESH10. Let $C1$ range over all cubes in R and $C2$ range over all cubes in T; then repeat RESH2–8.

Remarks

The reordering of RESH1 puts more "splitable" cubes at the top of the list. RESH2 and RESH3 initiate the pair-wise comparison loop. Conditions of RESH4 or RESH5, which result in the removal of a cube, may occur as a result of the current reshape process or as a result of a previous reduce process. RESH10 gives "stubborn" cubes another chance to be reshaped.

Discussion

• Summary

A general two-level logic function minimization technique, MINI, has been described. The MINI process does not generate all prime implicants nor perform the covering step required in a classical two-level minimization. Rather, the process uses a heuristic approach that obtains a near minimal solution in a manner which is efficient in both computing time and storage space.

MINI is based on the positional cube notation in which groups of inputs and the outputs form separate coordinates. Regular Boolean minimization problems are handled as a particular case. The capability of handling multiple output functions is implicit.

Given the initial specification and the partition of the variables, the process first maps or decodes all of the implicants into the cube notation. These cubes are then "exploded" into disjoint cubes which are merged, reshaped, and purged of redundancy to yield consecutively smaller solutions. The process makes rigorous many of the heuristics that one might use in minimizing with a Karnaugh map.

The main subprocesses are

1. Disjoint sharp.
2. Cube expansion.
3. Cube reduction.
4. Cube reshaping.

The expansion, reduction, and reshaping processes appear to be conceptually new and effective tools in practical minimization approaches.

• Performance

The MINI technique is intended for "shallow" functions, even though many "deep" functions can be minimized successfully. The class of functions which can be minimized is those whose final solutions can be expressed in a few hundred cubes. Thus, the ability to minimize a function is not dependent on the number of input variables or minterms in the function. We have successfully minimized several 30-input, 40-output functions with millions of minterms, but have failed (due to the storage limitation of an APL 64-kilobyte work space) to minimize the 16-variable EXCLUSIVE OR function which must have 2^{15} cubes in the final solution.

For an n-input, k-output function, define the effective number of input variables as $n + \log_2 k$. For a large class of problems, our experience with the APL program in a 64-kilobyte work space indicates that the program can handle almost all problems with 20 to 30 effective inputs. The number of minterms in the problem is not the main limiting factor.

The performance of MINI must be evelauted using two criteria. One is the minimality of the solution and the other is the computation time. Numerous problems with up to 36 effective inputs have been run; MINI obtained the actual minimum solution in most of these cases. The symmetric function of nine variables, S_{3456}^9, contains 420 minterms and 1680 prime implicants when each variable is in its own part (i.e., the regular Boolean case). The minimum two-level solution is 84 cubes. The program produced an 85-cube solution in about 20 minutes of 360/75 (APL) CPU time. The minimality of the algorithm is thus shown to be very good, considering the difficulty of minimizing symmetric functions in the classical approach, due to many branchings. A large number of very shallow test cases, generated by the method shown in [19], were successfully minimized, although a few cases resulted in one or two cubes more than the known minimum solutions.

The run time is largely dependent on the number of cubes in the final solution. This dependence results because the number of basic operations for the expand, reduce, and reshape processes is proportional to the square of the number of cubes in the list. It is difficult to compare the computation time of MINI with classical approach programs. The many-variable problems run on MINI could not be handled by the classical approach because of memory space and time limitations. For just a few input variables (say, up to eight variables), both approaches use comparable run times. However, the complexity of computation grows more or less exponentially with the number of variables in the classical minimization, even though the problem may be a shallow one. An assembly language version of MINI is now almost complete. The run time can be reduced by a factor as large as 50, requiring only a few minutes for most of the 20- to 30-effective-input problems. Thus it appears that MINI is a viable alternative to the classical approach in minimizing the practical problems with many input and output variables.

• Minimal solutions in the classical sense

The MINI process tries to minimize the number of cubes or implicants in the solution. The cubes in the solution may not be prime, as in the classical minimization where the cost function includes the price (number of input connections to AND and OR gates) of realizing each cube. But if such consideration becomes beneficial, a prime cube solution *can* be obtained from the result of MINI. This is done by first applying the reduction process to the output part of each cube in the solution and then expanding all the input parts of the cubes in any arbitrary part order. The MINI solution can also be reduced to smaller cubes by putting through an additional reduction step.

• *Multiple-valued logic functions*
It was mentioned that each part of the generalized universe may be considered as a multiple-valued logical input. By placing n_i Boolean variables in part i, we presented the MINI procedure with part lengths equal to 2^{n_i} except for the output part. MINI can handle a larger class of problems if the specification of a function is given directly in the cube format.

By organizing problems such as medical diagnoses, information retrieval conditions, criteria for complex decisions, etc. in multiple-value variable logic functions, one can minimize them with MINI and obtain aid in analysis. This is demonstrated with the following example.

Example: Nim
The game of *Nim* is played by two persons. There is an arbitrary number of piles of matches and each pile may initially contain an arbitrary number of matches. Each player alternately removes any number (greater than zero) of matches from one pile of his choice. The player who removes the last match wins the game. The strategy of the game has been completely analyzed and the player who leaves the so-called "correct position" is assured of winning the game, for the other player must return it to an incorrect position, which can then be made into a correct one by the winning player.

The problem considered contains five piles and each pile has two places for matches. Thus a pile can have no match, one match, or two matches at any phase of the game. Taking the number of matches in a pile as values of variables, we have a five-variable problem and each variable has three values (0, 1 or 2). Out of 243 (3^5) possible positions, 61 are correct. The 182 remaining incorrect positions were specified and minimized by the MINI program. For instance, the incorrect position (0, 1, 1, 0, 2) is specified as (100 010 010 100 001) in the generalized coordinate format. Using this result and the fact that all variables are symmetric, one can deduce the incorrect positions:

1. Exactly two piles are empty (cubes 1 – 10) or no pile is empty (cube 21).
2. Only one pile has two matches (cubes 11, 13, 17 – 19).
3. Only one pile has one match (cubes 12, 14 – 16, 20).

The MINI result identified the 21 cubes shown below.

1	011	100	011	100	011
2	100	011	100	011	011
3	100	011	011	100	011
4	011	011	100	011	100
5	011	100	011	011	100
6	011	100	100	011	011
7	011	011	011	100	100
8	100	011	011	011	100
9	011	011	100	100	011
10	100	100	011	011	011
11	001	110	110	110	110
12	101	101	101	010	101
13	110	110	110	001	110
14	010	101	101	101	101
15	101	010	101	101	101
16	101	101	010	101	101
17	110	001	110	110	110
18	110	110	001	110	110
19	110	110	110	110	001
20	101	101	101	101	010
21	011	011	011	011	011

• *Further comments*
In the case of the single output function F, the designer invariably has the option of realizing either F or $\overline{F}$. The freedom to realize either is often a consequence of the availability of both the true and the complemented outputs from the final gate. However, it may also be a consequence of the acceptability of either form as input to the next level. Given the choice of the output phases for a multiple-output function, a best phase assignment would be the one that produces the smallest minimized result. Since there are 2^k different phase assignments for k output functions, a non-exhaustive heuristic method is desired. One way to accomplish this would be to double the outputs of the function by adding the complementary phase of each output before minimization. The phases can be selected in a judicious way from the combined minimized result. This approach adds only a double-size output part in the MINI process. The combined result is just about double the given one-phase minimization. Hence, using the MINI approach a phase-assigned solution can be attained in about four times the time required to minimize the given function that has every output in true phase.

Our successful experience with the MINI process suggests both challenging theoretical problems and interesting practical programs. A theoretical characterization of functions, which either confirms or refutes the MINI heuristics, would be useful. The number of times the cube expansion process need be iterated is another matter requiring further study. Currently, we terminate the iteration if there is no improvement from the previous application of the expansion step.

Acknowledgments

We thank Harold Fleisher and Leon I. Maissel for their constructive criticism and many useful discussions in the course of the development of MINI. James Chow has implemented most of it in assembly language with vigor and originality.

Appendix

Here we give an example of a four-input, two-output Boolean function to illustrate the major steps of MINI. Most functions of this size get to the minimal solution by the first expansion alone. This example, however, is an exception and illustrates all the subprocesses of MINI. We use the Karnaugh map for the illustrations, rather than the cube notation. The conventions for this map are as follows:

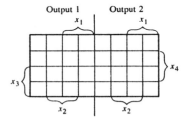

The ordered cubes are denoted by numbers in the vertices of the cubes, as follows.

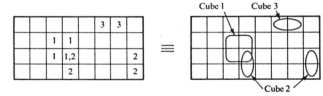

The function to be minimized and the effects of the subprocesses are shown in Fig. A1.

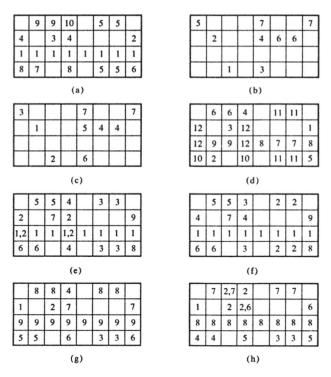

Figure A1 Example of the major steps of MINI operating on a 23-minterm list.

(a) Unit-distance-merged F, ordered for ⊛ right argument.

(b) Disjoint $\overline{F}$ ordered for expansion against F of a.

(c) Expanded $\overline{F}$ ordered for ⊛ right argument.

(d) Disjoint F ordered for expansion against $\overline{F}$ of c; beginning of the decreasing solution.

(e) Expanded F ordered for reduction.

(f) Reduced F ordered for reshaping.

(g) Reshaped F ordered for another expansion against $\overline{F}$.

(h) Final expanded F, the minimal solution of eight cubes: cubes 1, 3 – 6 are not prime.

References

1. W. V. Quine, "The Problem of Simplifying Truth Functions," *Am. Math. Monthly* **59**, 521 (1952).
2. W. V. Quine, "A Way to Simplify Truth Functions," *Am. Math. Monthly* **62**, 627 (1955).
3. E. J. McCluskey, Jr., "Minimization of Boolean Functions," *Bell Syst. Tech. J.* **35**, 1417 (1956).
4. J. P. Roth, "A Calculus and An Algorithm for the Multiple-Output 2-Level Minimization Problem," Research Report RC 2007, IBM Thomas J. Watson Research Center, Yorktown Heights, New York, February 1968.
5. E. Morreale, "Recursive Operators for Prime Implicant and Irredundant Normal Form Determination," *IEEE Trans. Comput.* **C-19**, 504 (1970).
6. J. R. Slagle, C. L. Chang and R. C. T. Lee, "A New Algorithm for Generating Prime Implicants," *IEEE Trans. Comput.* **C-19**, 304 (1970).
7. R. E. Miller, *Switching Theory, Vol. 1: Combinatorial Circuits*, John Wiley & Sons, Inc., New York, 1965.
8. E. Morreale, "Partitioned List Algorithms for Prime Implicant Determination from Canonical Forms," *IEEE Trans. Comput.* **C-16**, 611 (1967).
9. C. C. Carroll, "Fast Algorithm for Boolean Function Minimization," Project Themis Report AD680305, Auburn University, Auburn, Alabama, 1968 (for Army Missile Command, Huntsville, Alabama).
10. R. M. Bowman and E. S. McVey, "A Method for the Fast Approximation Solution of Large Prime Implicant Charts," *IEEE Trans. Comput.* **C-19**, 169 (1970).
11. E. G. Wagner, "An Axiomatic Treatment of Roth's Extraction Algorithm," Research Report RC 2205, IBM Thomas J. Watson Research Center, Yorktown Heights, New York, September 1968.

12. Y. H. Su and D. L. Dietmeyer, "Computer Reduction of Two-Level, Multiple Output Switching Circuits," *IEEE Trans. Comput.* **C-18**, 58 (1969).
13. R. S. Michalski, "On the Quasi-Minimal Solution of the General Covering Problem," *Proceedings of the Fifth International Symposium on Information Processing (FCIP69)* **A3**, 125, 1969 (Yugoslavia).
14. R. S. Michalski and Z. Kulpa, "A System of Programs for the Synthesis of Combinatorial Switching Circuits Using the Method of Disjoint Stars," *Proceedings of International Federation of Information Processing Societies Congress 1971.* Booklet TA-2, p. 158, 1971 (Ljubljana, Yugoslavia).
15. D. L. Starner, R. O. Leighon and K. H. Hill, "A Fast Minimization Algorithm for 16 Variable Boolean Functions," submitted to *IEEE Trans. Comput.*
16. Y. H. Su and P. T. Cheung, "Computer Minimization of Multi-valued Switching Functions," *IEEE Trans. Comput.* **C-21**, 995 (1972).
17. J. P. Roth, "Theory of Cubical Complexes with Applications to Diagnosis and Algorithmic Description," Research Report RC 3675, IBM Thomas J. Watson Research Center, Yorktown Heights, New York, January 1972.
18. S. J. Hong and D. L. Ostapko, "On Complementation of Boolean Functions," *IEEE Trans. Comput.* **C-21**, 1072 (1972).
19. D. L. Ostapko and S. J. Hong, "Generating Test Examples for Heuristic Boolean Minimization," *IBM J. Res. Develop.* **18**, 469 (1974); this issue.

Received December 10, 1973; revised April 30, 1974

S. J. Hong, a member of the IBM System Products Division laboratory in Poughkeepsie, New York, is on temporary assignment at the University of Illinois, Urbana, Illinois 61801; R. G. Cain is located at the IBM System Development Division Laboratory, Poughkeepsie, New York 12602; and D. L. Ostapko, is located at the IBM System Products Division Laboratory, Poughkeepsie, New York 12602.

ESPRESSO-MV: ALGORITHMS FOR MULTIPLE-VALUED LOGIC MINIMIZATION

R.L. Rudell
A.L.M. Sangiovanni-Vincentelli
Department of Electrical Engineering
and Computer Sciences
University of California
Berkeley. CA 94720

Abstract

We present an extension of the Espresso-II algorithms that efficiently performs multiple-valued logic minimization of large generalized Boolean functions. Results from many industry and University chip designs demonstrate the effectiveness of the algorithms. A lower-bound technique has been developed which proves that, for many cases, the algorithms actually produce the minimum solution.

1. Introduction

Espresso-II is a program for minimization of two-level binary-valued switching functions [1]. It was originally written in APL on an IBM 3033. Espresso-IIC [8] is a C-language implementation of the Espresso-II algorithms with several extensions to improve both the quality of the results, and the performance of the algorithms. Espresso-IIC has successfully minimized functions as large as 34 inputs and 30 outputs from 3280 terms down to 375 terms in 30 minutes on an IBM 3081. We believe Espresso-IIC to be a good solution to the problem of minimizing large switching functions.

We soon became interested in extending the Espresso-II algorithms to the general framework of multiple-valued variables. Multiple-valued logic has many uses in optimizing structures built from binary-valued logic. For example, it has been shown that the *input-encoding problem* can be solved by viewing the problem as a multiple-valued minimization problem. This can be applied to the optimal state-assignment problem (for many types of finite-state machines) [4], or to optimal assignment of opcodes in a microprocessor to reduce the instruction decode logic [4]. Multiple-valued logic functions can also be used to represent and minimize PLA's with *two-bit decoders* [2,3].

With an appropriate transformation, a multiple-valued minimization problem can be solved with any binary-valued minimizer [1]. However, this approach can be inefficient; the transformed problem has a don't-care set which can become excessively large, and the number of binary variables needed is equal to the sum of the number of multiple-valued parts in the original problem. Hence, even Espresso-IIC was unable to minimize the transformed function resulting from attempting to perform a state-assignment on a dense 93-state machine. (It should be noted that the transformed function had over 100 input variables and over 100 output functions, and there were more than 5000 don't-care terms.)

In this paper we present a set of algorithms for the minimization of multiple-valued functions. In particular, we extend the key concepts of Espresso-II for standard logic minimization to the multiple-valued case. We report experimental results that show the effectiveness of the algorithms. Interestingly, the Espresso-MV algorithms, when applied to standard logic minimization problems, produce solutions up to three times faster than the algorithms used in Espresso-II.

2. Statement of the Problem

A generalized Boolean function [2] of r multiple-valued variables is a function defined as:

$$f(x_1, \cdots, x_r): B^{n_1} \times \cdots \times B^{n_r} \to \{0,1,2\}$$

where $B \equiv \{0,1\}$ and 2 corresponds to a value function which can be either 0 or 1 (i.e., a *don't-care*). Each variable x_i is said to have n_i parts. We denote the parts of variable x_j as $x_j^1, x_j^2, ..., x_j^{n_j}$. (In the binary-valued case it is customary to denote the parts as $x_j \equiv x_j^1$ and $\bar{x}_j \equiv x_j^2$).

An n-input, m-output switching function can be represented by a generalized Boolean function [2] where $r = n + 1$; $n_i = 2, i = 1 .. n$; and $n_r = m$. Thus, the standard logic minimization problem for multiple-output functions can be seen as a multiple-valued minimization problem.

An expression can be written for a generalized Boolean function as a sum of *terms* (also called *implicants*), where each term is a product of *literals*. Each literal is a sum over a subset of the parts for a variable. This is called the sum-of-products form. (*Sum* and *product* are to be interpreted in the Boolean sense.) A useful representation of the expression for a function is the positional cube notation [6]. We introduce the positional cube by way of an example:

Example 1: A 3-variable function where the first variable has 3 parts, and the second and third variables have 4 parts:

$$F = (x_2^1 + x_2^2 + x_2^3)(x_3^3 + x_3^4)$$
$$+ (x_1^2 + x_1^3)(x_2^1 + x_2^2 + x_2^3)(x_3^3 + x_3^4)$$
$$+ (x_1^1 + x_1^3)$$

We can represent this expression in positional notation as:

x_1	x_2	x_3
111	1110	0011
011	1110	0011
101	1111	1111

Note that each row (called a *cube*) represents a term, and a 1 indicates the inclusion of the particular part of the variable in the term. A collection of cubes is referred to as a cover of the function.

There are many ways of writing an expression for a function and for each there is a unique positional cube representation. Our minimization problem is to find the expression which has the fewest number of terms (equivalently, the positional cube representation with the fewest number of rows).

2.1. Input Encoding Problem

The problem can be stated as follows: we are given a set of symbols $S \equiv \{s_1, s_2, \cdots, s_p\}$ and a Boolean function defined as

$$f: \{0,1\}^n \times S \to \{0,1,2\}^m$$

We seek an encoding of the symbols into binary vectors that minimizes the number of terms needed to represent the function. It has been shown [2] that this problem can be solved by performing a multiple-valued minimization of the function f (where S is represented by a single multiple-valued variable with p parts), and then solving an encoding problem which maps the result of the multiple-valued minimization into binary vectors for each symbol. This has been used as an approximation to the state-assignment problem [4] where the set S is the set of states, and the function f defines the output function as a function of the binary inputs and the present state.

Reprinted from *IEEE Proc. 1985 Cust. Int. Circ. Conf.*, pp. 230–234, May 1985.

2.2. Input Decoders

Four adjacent columns in a PLA arising from two input signals (say a and b) normally carry the logic values $a, \bar{a}, b, \bar{b}$. However, it is possible to pair two input variables and use a decoder in the input buffer of the PLA so that four adjacent columns carry the values $a+b$, $a+\bar{b}$, $\bar{a}+b$, and $\bar{a}+\bar{b}$. The core of the PLA with the decoders will have the same or fewer terms than the original PLA. However, the decoders allow several functions to be realized in a single product term (e.g., $a+b$, and $a\bar{b}+\bar{a}b$) which can sometimes significantly reduce the number of terms in the PLA. In many cases, the delay of the input decoders is negligible with respect to the delay incurred in the core of the PLA.

Once a pairing of the input variables has been chosen, it can be shown that minimization with the input decoders is a multiple-valued minimization problem [2]. This is done by viewing the pair of variables (a,b) as a multiple-valued variable with 4 parts.

3. The Espresso-MV Algorithms

Our goal is to take a generalized Boolean function and produce the minimum sum-of-products form. We briefly comment here that the complexity of extracting the minimum sum-of-products form is extremely high. A robust algorithm for attacking this problem must resort to heuristics to always provide a solution within a reasonable consumption of resources.

Figure 1 shows the algorithm used by Espresso-MV to solve this minimization problem. We refer the reader to [1] for definitions of the specialized terms which follow. F is the cover of the *ON-set* (those terms which imply the function is 1), R is the cover of the *OFF-set* (those terms which imply the function is 0), and D is the cover of the *Don't-care set* (those terms which imply the function can be either 0 or 1). The constituent algorithms are: computing the complement of a function (COMPLEMENT); expanding an implicant into a desirable *prime implicant* (EXPAND); extracting from a set of prime implicants a minimal, irredundant subset (IRREDUNDANT); detecting which of the prime implicants are *essential prime implicants* (ESSENTIAL); and reducing an implicant into the smallest implicant which, together with the remaining implicants, still covers the original function (REDUCE). The routines EXPAND, IRREDUNDANT, and REDUCE are iterated until no new implicants are removed from the cover.

A last attempt is made at reducing the function further by checking if there are any better solutions *close* to the current solution. All of the terms are reduced without replacement (REDUCE_GASP), and then expanded against each other to see if any term can expand to cover the reduced form of another term. If so, we have reason to believe that the inclusion of these terms will lead to a better solution.

Finally, the cover F is made maximally sparse by removing parts which are not required in the representation of the function.

The external view of the minimization strategy of Espresso-MV is identical to Espresso-IIC; however, each step has changed internally to accommodate multiple-valued variables. We will discuss below the algorithms and the modifications necessary for the multiple-valued case.

3.1. Shannon Expansion

The Shannon expansion [5] is used extensively in Espresso-II to perform several basic operations on a function. Recall that in the binary-valued case it is possible to partition a function F with respect to a particular variable x_j as:

$$F = x_j F_{x_j} + \bar{x}_j F_{\bar{x}_j}$$

where the cofactors of F (F_{x_j} and $F_{\bar{x}_j}$) are simpler functions in that they contain the same or fewer terms, and they depend on one less variable. For example, to compute the complement of a function (which is used by EXPAND), we recognize that

$$\bar{F} = x_j \bar{F}_{x_j} + \bar{x}_j \bar{F}_{\bar{x}_j}$$

To determine if a function is a tautology (i.e., to check if, for all values of the inputs, the function is 1), (which is used by IRREDUNDANT and ESSENTIAL) it is sufficient to check that F_{x_j} and $F_{\bar{x}_j}$ are both tautologies. To find the smallest term which contains the complement of a function (referred to as SCCC and used by REDUCE), we

find the smallest term which contains the complement of each cofactor, and then find the smallest term which contains these two terms.

Thus, the computational usefulness of the Shannon cofactor is obvious: to compute an operation on a function F, we recursively compute the operation for each cofactor F_{x_j} and $F_{\bar{x}_j}$ and merge the results to complete the operation for F. This is made efficient in Espresso-II by observing that there exists a simple class of functions for which the basic operations mentioned above can be computed efficiently. These are referred to as *unate* functions [1]. Therefore, to proceed from the binary-valued case to the multiple-valued case, we seek a generalization of the Shannon expansion, and a generalization of the concept of a unate function. We would like to find the extensions of these concepts which will also preserve the desirable properties that we find in the binary-valued case.

One obvious generalization of the Shannon expansion is:

$$F = \sum_{i=1}^{n_j} x_j^i F_{x_j^i} \qquad (1)$$

which partitions the function on variable x_j into n_j cofactors. Each of the cofactors $F_{x_j^i}$ is independent of the variable x_j, and has the same or fewer terms than F. Presumably we can compute the basic operation on each of these simpler functions, and then perform an n_j-way merge of the results.

Another possible generalization is:

$$F = l_j F_{l_j} + r_j F_{r_j} \qquad (2)$$

where l_j and r_j are literals which form a partition of the parts in variable x_j (i.e., $l_j + r_j = 1$ and $l_j r_j = \phi$). We refer to F_{l_j} as the left partition, and F_{r_j} as the right partition of the function. This allows a binary recursion and a binary merge to be used in place of the n_j-way recursion and the n_j-way merge. However, it does not necessarily eliminate the entire variable x_j — it merely eliminates the parts of r_j from F_{l_j} and the parts of l_j from F_{r_j}. (Note that in the binary-valued case, equations (1) and (2) are equivalent.)

For efficiency, we seek the method which allows the greatest degree of freedom in choosing how to partition the function. Note that in the first case our only choice is in the splitting variable. Once that is chosen, we proceed to completely eliminate that variable, and then proceed to the next variable. Equation (2) allows us to select both the splitting variable and the parts which are to belong to each of l and r. Further, after we split the function into two halves with respect to the variable, we have, at the next step in the recursion, the freedom to choose either the same or a different variable. Often this can lead to a solution with fewer levels of recursion. It should be clear that equation (2) reduces to equation (1) if the same variable is chosen as the splitting variable until it is completely eliminated.

3.2. Choice of splitting variable

The choice of the splitting variable strongly determines the performance of the algorithms. As mentioned earlier, Espresso-II showed showed that the basic operations described above (complement, tautology, and SCCC) are easily computed for unate functions. Therefore, we first generalize the concept of a unate function.

Definition: A multiple-valued function is monotone increasing (decreasing) in a part of a variable x_j, say x_j^i, if setting $x_j^i = 1$ (equivalently setting x_j to have its ith part a 1 and all other parts 0) causes the function, if it changes, to increase (decrease) from 0 to 1 (from 1 to 0).

A multiple-valued logic function which is monotone increasing or decreasing in all the parts of variable x_j is said to be monotone or *unate* in x_j. A function which is monotone in all of its variables is said to be a unate function.

Looking at example 1, we see that f is monotone increasing in x_1^3, x_2^1, x_2^2, x_3^3, and x_3^3, and f is monotone decreasing in x_2^4, x_3^2, and x_3^2. f is not monotone in x_1^1 or x_1^2. In fact, changing x_1 from x_1^1 to x_1^2 may cause the function to increase or decrease according to the values of the other variables. Hence, f is unate in variables x_2 and x_3, but not in x_1.

To detect whether a specific cover of a function is unate in x_j, we simply check whether all the columns corresponding to x_j which are not all 1 are identical. In this case, the function is monotone increasing in the parts corresponding to columns which are all 1, and monotone decreasing in the parts corresponding to the columns identified as identical.

With this definition, we can prove the generalizations of the Espresso-II results for unate functions. For example, the following results simplify the tautology and SCCC operations:

Theorem (unate reduction): A function F is a tautology if and only if the reduced function which arises from considering only those terms which have all 1's in all unate variables is also a tautology.

Corollary: A unate function is a tautology if and only if some term in the function is the universal term (i.e., a term which is all 1's).

Theorem (SCCCU): The smallest term containing the complement of a unate function is the intersection of the smallest term containing the complement of each individual term.

These theorems (which are proved in [7]) provide us with an efficient termination for our recursion. Therefore, we are motivated to choose the splitting variable and the partition of that variable so as to eliminate variables which are the *least* unate. At the same time, we want to avoid (as much as possible) duplication of terms in the left and right partitions.

Algorithm 1 (Choice of splitting variable): for each variable, examine the columns which are not all 1's (i.e., the active parts), and determine the number of different columns. Choose to split on the variable with the greatest number of different columns. Break ties by considering, in order, the variable with the fewest cubes which have a full part (i.e., all 1's) in the variable, and the variable with the smallest deviation from a uniform distribution of ones over the columns.

Algorithm 2 (Choice of the partition of the active parts): Let m be the largest number of identical columns in the splitting variable. If $m > 1$, place the parts corresponding to the first m identical columns into the left partition, and the remaining parts in the right partition. If $m = 1$, place the first $n/2$ active parts into the left partition, and place the remaining $n/2$ active parts into the right partition.

More advanced heuristics can be imagined for the choice of the partition. For example, one goal might be to partition the active variables so as to keep the sum of the sizes of the left and right partitions small while making them as nearly equal in size as possible. This is to avoid duplicating terms in the left and right partitions. Unfortunately, the cost of computing a better partition must be weighed against its possible benefits. The simple strategy outlined here is very simple to compute, and hence was our first choice for experimentation. We are currently investigating other algorithms for the choice of partition of the splitting variable.

3.3. COMPLEMENT

We use the following to compute the complement:

$$\overline{F} = l_j \overline{F}_{l_j} + r_j \overline{F}_{r_j}$$

While there are variables which are not unate, we use the strategy described above to choose the partitions of the splitting variable. When the function becomes unate, we use an alternate strategy which finds the cube with the fewest active variables and then eliminates one of these variables. The recursion ends when there is only a single term at a leaf (at which point De Morgan's theorem is applied to compute the complement of a single term). Merging the results is done by checking for terms in $\overline{F}_{l_j}$ and $\overline{F}_{r_j}$ which are distance-1 apart (in which case the splitting variable can be expanded, and one term can be deleted). We also check each term of $\overline{F}_{l_j}$ to see if it is single-term contained by a term of $\overline{F}_{r_j}$, and vice-versa (in which case the splitting variable can be lifted). This often leads to a smaller representation of the complement.

For binary functions, Espresso-MV does not treat the output multiple-valued part any differently than any other multiple-valued part. Thus, COMPLEMENT computes a more convenient representation of the complement (as opposed to the single-output complementation strategy of Espresso-II).

3.4. EXPAND

The expansion of an implicant c proceeds according to the following heuristics: (1) If there are any other implicants of F which can be contained by an expansion of c, then c will first expand as necessary to cover as many implicants as possible; (2) If there are no implicants which can be contained by an expansion of c, then, in a manner similar to MINI [6], the expansion direction is chosen by examining which direction will overlap the most other terms. MINI used a static ordering where, before any expansion was performed, a weight was computed and then used to rank the order in which expansion would be attempted. Espresso-MV uses the same weighting function but with a dynamic ordering (i.e., the directions are ranked after the expansion of each part); (3) Finally, when there are no terms which can be overlapped, we attempt to form the largest possible prime. This is done by mapping the expansion into a set-covering problem which is then solved with a heuristic algorithm.

Before applying any of the steps above, we first remove from consideration directions which cannot be expanded, and expand in directions which can always be expanded without affecting further expansions.

Espresso-II used the complement in an *unwrapped* form in which each term in the complement was required to have only a single part in the output variable. This facilitated the description of the expansion process as a covering problem. Generalizing the EXPAND operation in this way would require each multiple-valued variable to have only a single part. This is unsuitable as it could require an explosion in the number of terms needed to represent the complement. Therefore, heuristics (1) and (2) have been restated in terms of *distance* which easily generalizes to the multiple-valued case. (For example, if there is a term in the complement which is distance 1 from c, then we cannot expand in this direction; or, we can expand to cover another term if and only if the term resulting from the expansion is always greater than zero distance from the complement.) Only when attempting to find the largest possible prime implicant (step 3) are we forced to unravel the remaining part of the complement, and this need only be done when the unravel results in a reasonable sized covering problem. More details can be found in [7].

The EXPAND algorithm makes efficient use of the complement. The heuristics used in the expansion would be prohibitively expensive if the complement of the function were not computed first. Also, Espresso-MV treats all of the multiple-valued parts equally. Therefore, the implicant is expanded to a multiple-valued prime implicant.

3.5. IRREDUNDANT

Once every implicant has been expanded into a prime implicant, our goal is to extract a minimal subcover of these prime implicants to form an irredundant representation of the function. The simplest technique for this problem is to merely pick a prime implicant, decide if it is essential or redundant, and discard the prime implicant if it is redundant. Unfortunately, this simple technique fails to intelligently break any cyclic dependencies, and in the worst case, may be far from the minimal solution. Therefore, we choose to expend more effort in an attempt to extract a minimal subcover.

We first split F into the relatively-essential prime implicants (E), and the redundant prime implicants (R). By definition, the implicants of E contain some minterm which is not covered by any of the other primes in F, and hence all of E must be in any minimal subcover of F. The implicants in R are further divided into the totally redundant subset (R_t) and the partially redundant subset (R_p). The implicants of R_t are those implicants which are contained by E, and hence they can never be in a minimum subcover of F.

To decide whether an implicant of F belongs to either E or R, or whether a implicant of R belongs to either R_t or R_p requires a check that a function *covers* the given implicant. This is equivalent [1] to checking whether the function F_c (read F cofactored against c) is a tautology. Thus, tautology checking is a basic operation needed for the IRREDUNDANT algorithm.

The implicants of R_p form cyclic redundancies. In order to select a good subset of these primes, we use a recursive procedure similar to TAUTOLOGY to map the problem into a set-covering problem. The set-covering problem is solved by first finding a large (not necessarily maximum) group of the sets which are all independent (i.e., they have no element in common). It is obvious that at least one element from each of these sets must belong to the minimum cover, and hence the size of this group of sets is a lower bound on the size of the minimum cover. Further, the heuristic selection of elements can start from the independent set.

3.6. REDUCE

We now have a prime and irredundant cover, which is a local minimum of our cost function. Unfortunately, it may be far from the global minimum, and hence we choose to iterate to avoid being stuck in a local minimum. This is done by shrinking each implicant back to the smallest implicant possible while not changing the function. The smallest term containing the complement of the function without the term (SCCC) is computed recursively. The maximal reduction of the term is the intersection of the term and the SCCC.

4. Results

4.1. Binary Valued Functions

Tables 1 and 2 show the extent to which the algorithms have been tested. A detailed description of each example and results will appear in [7]. More than half of the 56 examples in Table 1 were collected from industrial and University chip designs, and the remaining examples are standard functions (e.g., addition, multiplication, square root) and known difficult examples (symmetric functions). Table 2 are industrial examples provided by H. De Man of the University of Leuven.

It is interesting to note that in Table 1 the solutions provided are similar to the results published for Espresso-IIC, but that the total time has been reduced by almost 50% (some of the larger examples actually ran up to three times faster). This is a surprising result, as one might expect the generalization of the algorithms to multiple-valued variables to penalize the performance for binary-valued minimization problems. However, the algorithms are actually improved by a uniform treatment of the multiple-output part of the function as a single multiple-valued variable. Hence, the Shannon cofactor is applied to the output multiple-valued part, and the heuristics used for selection of the splitting variable are allowed to uniformly consider the output part.

4.2. Results on Multiple Valued Functions

Table 3 show compares the effectiveness of Espresso-MV against mapping the multiple-valued problem into an equivalent binary-valued minimization. For DK16 (a 27-state finite-state machine), the transformed binary-valued minimization took 108.6 seconds on the IBM 3081, while the direct multiple-valued minimization took only 1.6 seconds. The results provided in each case where the same. Another notable result is that the original problem of state-assignment for the dense 93-state machine was performed rather routinely (reduced from over 3000 terms to 770 terms) in 17.5 minutes on an IBM 3081.

Example MULT4 and PCC.FSM are also shown as an example of using two-bit decoders on the input of the PLA. Note that without the decoders, MULT4 requires 127 product terms, and PCC.FSM requires 57 product terms. The binary-valued minimization of PCC.FSM could not be performed in 30 minutes on the IBM 3081, but the multiple-valued minimization required only 4.4 seconds.

4.3. Quality of the results

Perhaps too often heuristic techniques are presented and claimed optimal with little theoretical or experimental evidence to support this claim. Therefore, we undertook an experiment to substantiate our claim that Espresso-MV produces a high quality result.

The Espresso-MV IRREDUNDANT algorithm as described above is very powerful — it extracts a minimal cover, and also computes a lower bound on the size of the minimum cover. For some minimization problems, it is possible to generate the set of all prime implicants and then apply the IRREDUNDANT strategy to this set thus providing a lower bound on *any* minimal solution. If Espresso-MV is able to find a solution which achieves this lower bound, then we have a proof that Espresso-MV has achieved the minimum solution. The

generation of all of the prime implicants is made efficient with an algorithm described elsewhere [7], and the covering table is efficiently generated using the *unate recursive paradigm*. The final covering is then done heuristically, and is also quite fast.

This was successful for 54 of the 81 functions (67 %) reported in Tables 1 and 2. In 40 of these 54 cases (74 %), Espresso-MV reached a solution which equaled the lower bound, and hence has achieved the minimum solution (even though the minimum solution was sometimes intractable by direct methods), For the remaining 14 cases, Espresso-MV was usually within 3 product terms from the lower bound, and in the worst case was 12 % from the best lower bound that could be determined. (In fact, Espresso-MV may have the minimum solution for some of these cases, but we cannot prove it.) It is important to note that placing the lower-bound on the function size was anywhere from 10 to 100 times as expensive as minimizing the function with Espresso-MV. The advantage of Espresso-MV is that even in those cases where we can't generate the set of all prime implicants, we still receive a solution of presumably high quality.

4.4. Availability

Espresso-MV has been run on a variety of machines including the VAX 11/780 (Unix and VMS), the IBM 3081 (VM/CMS), the IBM PC (PC-DOS), and the SUN workstation (68000-based Unix machine). The source code and all of the PLA examples mentioned here are available on tape from the EECS Industrial Liaison Program, 457 Cory Hall, University of California, Berkeley, California, 94720.

References

[1] R. Brayton, G.D. Hachtel, C. McMullen and A.L. Sangiovanni-Vincentelli, *"Logic Minimization Algorithms for VLSI Synthesis"*, Kluwer Publishing Co., Boston, 1984

[2] T. Sasao, "Multiple-valued decomposition of generalized Boolean functions and the complexity of programmable logic arrays", *IEEE Trans. on Comp.*, Vol C-30, No. 9, pp. 635-643, September, 1981.

[3] H. Fleisher and L.I. Maissel, "An Introduction to array logic," *IBM J. Res. and Dev.*, Vol 19., pp. 98-108, March 1975.

[4] G. De Micheli, *Computer-Aided Synthesis of PLA-Based Systems*, PhD Dissertation, University of California, 1983. (Chapter 4).

[5] C.E. Shannon, "The synthesis of two-terminal switching circuits," *Bell Sys. Tech. J.*, 1948

[6] S.J. Hong, R.G. Cain, and D.L. Ostapko, "MINI: A heuristic approach for logic minimization," *IBM J. of Res. and Dev.*, Vol 18, pp 443-458, September 1974.

[7] R.L. Rudell, Master's Project Report, in preparation.

[8] R.L. Rudell, "Espresso-IIC User's Manual", *CAD Toolbox User's Manual*, U.C. Berkeley.

```
/*
espresso — minimize a multiple-valued Boolean function
    F refers to the ON-set of the function
    D refers to the don't-care set of the function
    R refers to the OFF-set of the function
*/
espresso(F, D)
{
    /* Compute the complement of the function */
    R ← COMPLEMENT(F + D);

    /* Initial expand and irredundant */
    F ← EXPAND(F, R);
    F ← IRREDUNDANT(F, D);

    /* Remove essential primes from F */
    E ← ESSENTIAL(F, D);
    F ← F - E;
    D ← D + E;

    do {
        φ₂ ← cost(F);

        /* Repeat inner loop until solution becomes stable */
        do {
            φ₁ ← cost(F);
            F ← REDUCE(F, D);
            F ← EXPAND(F, R);
            F ← IRREDUNDANT(F, D);
        } while (cost(F) < φ₁);

        /* Perturb solution to see if we can continue to iterate */
        G ← REDUCE_GASP(F, D);
        G ← EXPAND(G, R);
        F ← IRREDUNDANT(F + G, D);

    } while (cost(F) < φ₂);

    /* Return the essential primes to F */
    F ← F + E;
    D ← D - E;

    /* Make the final solution sparse */
    F ← MAKE_SPARSE(F, D, R);

    return F;
}
```

Figure 1. The Espresso-MV main algorithm.

Program	Terms	Literals	Time
Espresso-MV	5993	60322	560 sec
Espresso-IIC	6001	60578	992 sec
Espresso-II	5998	60432	4002 sec
MINI	6032	60837	44441 sec

Table 1a. Results for test set 1
(56 PLAs reported in [1])
Times are for IBM 3081, VM/CMS.

Program	Terms	Literals	Time
Espresso-MV	5993	60322	6690 sec
Pop	6828	66442	7443 sec

Table 1b. Results for test set 1
(56 PLAs reported in [1])
Times are for VAX 11/780.

Program	Terms	Literals	Time
Espresso-MV	1812	17455	3528 sec
Prestol-II	1821		3594 sec

Table 2. Results for test set 2
(24 examples from the University of Leuven)
Times are for VAX 11/780.

Example *mi2* has been deleted due to a discrepancy in the result reported by Prestol-II and the lower bound which was determined for the function.

Example	in	out	Binary-valued		Multiple-valued	
			Terms	Time	Terms	Time
DK14	8	5	26	4.3 sec	26	0.5 sec
DK15	8	5	17	1.8 sec	17	0.3 sec
DK16	4	3	55	108.6 sec	55	1.6 sec
BLUE	18	14		3600 sec *	775	1053 sec
MULT4	16	8	89	38.1 sec	89	6.4 sec
PCC.FSM	32	25		1800 sec *	48	4.4 sec

Table 3. Comparison of direct multiple-valued minimization
to transforming the problem to an equivalent binary-valued
minimization (IBM 3081 version)

* computation did not terminate

McBOOLE: A New Procedure for Exact Logic Minimization

MICHEL R. DAGENAIS, STUDENT MEMBER, IEEE, VINOD K. AGARWAL, SENIOR MEMBER, IEEE, AND
NICHOLAS C. RUMIN, SENIOR MEMBER, IEEE

Abstract—A new logic minimization algorithm is presented. It finds a minimal cover for a multiple-output boolean function expressed as a list of cubes. A directed graph is used to speed up the selection of a minimal cover. Covering cycles are partitioned and branched independently to reduce greatly the branching depth. The resulting minimized list of cubes is guaranteed to be minimal in the sense that no cover with less cubes can exist. The *don't care* at output is handled properly. This algorithm was implemented in C language under UNIX BSD4.2. An extensive comparison with ESPRESSO IIC shows that the new algorithm is particularly attractive for functions with less than 20 input and 20 output variables.

I. INTRODUCTION

THE design of VLSI circuits has been automated to the point where, in particular, there are programs which read logical relations and convert them to an equivalent list of cubes that can realize the desired function. Then, layout assemblers take this list of cubes and produce a programmable logic array (PLA) circuit. One example of such a system was presented by Meyer *et al.* [12]. Many programs have been developed to reduce the number of cubes needed to represent a function and, accordingly, to reduce the chip area consumed by the PLA circuit.

Many years ago, Quine [14] and McCluskey [11] presented a procedure that will always lead to a cover with the minimal number of cubes for a given function. Later, Tison [20], Roth [17], and Dietmeyer [8], among others, modified this procedure to improve its computational requirements. Some programs have been reported lately that provide a minimal cover, based on the procedures mentioned earlier. Ishikawa [10] presented a minimizer that produces a minimal cover based on the procedure of Tison [20]. However, he mentions that it cannot handle, in reasonable time on a mainframe computer, functions having more than 10 input variables. Brayton *et al.* [6] also state that, with existing minimizers, it is impractical to produce a minimal cover for even a medium-sized function with 10 to 15 input variables.

On the other hand, many heuristic procedures have been developed and give good results in minimizing a list of

Manuscript received March 18, 1985; revised June 7, 1985. This work was supported by a Fellowship to the first author, and by a strategic grant, both from the Natural Science and Engineering Research Council of Canada.

The authors are with the Department of Electrical Engineering, McGill University, Montreal, P.Q., Canada H3A 2A7.

IEEE Log Number 8406324.

cubes. Even though these procedures often reach a solution which is a minimal cover, in general they give a near-minimal cover with no information on how far it is from a true minimal solution. These minimizers include PRESTO [7], PRESTOL-II [2], MINI II [17], and ESPRESSO II [5]. The latter and its more recent version, ESPRESSO-IIC, have been carefully evaluated [6]. They have been shown to give a very good minimal or near-minimal cover for fairly large functions, in a reasonable CPU time. Other programs such as CAMP [3] produce a minimal cover for medium sized functions when no covering cycles are present.

The new minimizer McBOOLE 21 is based on some very efficient graph and partitioning techniques, and it can find a minimal cover for larger functions than those that could be practically handled up to now. In this paper, the general theory to obtain a minimal cover is first briefly reviewed. The new procedure and structures implemented in McBOOLE are then detailed. Some results and comparisons of McBOOLE and ESPRESSO IIC are presented. Finally the performance and limitations of McBOOLE are discussed. It is assumed that the reader is familiar with the cubic notation for logic functions [8].

II. NOTATION

A multiple output boolean function of m inputs and n outputs is defined as follows:

$$\text{input space: } B = \{0, 1\}$$

$$\text{output space: } Y = \{0, 1, d\}$$

$$\text{function } f: B^m \rightarrow Y^n.$$

The value d (*don't care*) at output means that the value is unspecified, and a value of 0 or 1 will be accepted to realize this part of the function. Such a function can be represented by a list of cubes. Each cube contains an input part and an output part as follows:

$$\text{input part: } m \text{ literals that can be } \{0, 1, x\}$$

$$\text{output part: } n \text{ literals that can be } \{0, 1, d\}.$$

The input part identifies the portion of the input space to which a cube applies. The x in the input part matches all the points of the function that have either a 1 or a 0 for this variable. For example a cube with k x's in its input part applies to 2^k points of the function. In a typical truth

Reprinted from *IEEE Trans. CAD of Int. Circ. Syst.*, vol. CAD-5, no. 1, pp. 229–238, Jan. 1986.

75

table, all the cubes are disjoint and there is no ambiguity. In some cases however, the cubes at input are not disjoint, and many cubes do apply to a point of the function. Such a case usually arises when the cubes were produced by some program; the value of the output vector for this point has then to be determined by priority rules, consistent with the other programs. A d in any of the cubes overrides the 0 or 1 present in any other cube for the same output variable. Similarly, the 1 overrides the 0. When no cubes apply to a point of the function, the output vector is composed of all 0's by default.

For each point in the input space, the output variables having the value 1 are called the *minterms;* therefore, there can be up to n minterms for each of the 2^m points of the input space. Some useful operators are described below. They can apply to a list of cubes as well as to single cubes. Also, any set of minterms of a multiple-output logic function can be expressed as a list of cubes.

The *intersection* of two tubes, $c_1 \sqcap c_2$, is the set of minterms that belong to both c_1 and c_2. Two cubes are said not to intersect when their intersection is empty.

A cube c_1 *covers* another cube c_2, $c_2 \sqsubseteq c_1$, when all the minterms contained in c_2 are also contained in c_1. One can also say that c_2 is contained in c_1.

The *sharp* product of two cubes, $c_1 \# c_2$, is the set of minterms that belong to c_1 but not to c_2. One can also say that c_2 is subtracted from c_1.

The *star* product of two nonintersecting cubes, $c_1 * c_2$, produces a new cube c_3 which is the largest cube that can be formed from both c_1 and c_2.

III. The Extraction Algorithm

The procedure for reaching a minimal cover presented by Quine [14] and McCluskey [11] forms the basis of McBOOLE. Following is a summary of this procedure.

Suppose a given function is represented by a list of cubes F. Let PI be the list of *prime* cubes. A prime cube is any cube that satisfies the condition:

$$c \sqsubseteq F \text{ and for all possible cubes } c_i \sqsubseteq F$$
$$c \not\sqsubseteq c_i \text{ unless } c = c_i. \quad (1)$$

The procedure uses three lists. One list of cubes contains the *don't care* cubes and is named DC. A second list contains the prime cubes for which no decision has been reached as to whether they will be part of the minimal cover; it is the list of *undecided* cubes U. Finally, the cubes that are part of the minimal cover are stored in the list of *retained* cubes R. The procedure to extract a minimal cover from F is as follows:
1) Compute all the prime cubes of F and put them in the undecided list U. Place all the *don't care* cubes in the list DC.
2) Extract all the *extremal* prime cubes, remove them from U, and place them in the *retained* list R. An

extremal prime cube is defined as follows:

$$c_i \in U \text{ and } c_i \# (U \cup R \cup DC - c_i) \neq \emptyset \quad (2)$$

3) Delete from the list U all the *inferior* prime cubes for which the part not intersecting with $R \cup DC$ is contained in a single other cube of U. More formally, an inferior cube is defined as follows:

$$c \in U, \quad \exists c_i \in U, \quad c_i \neq c, \quad c\#(R \cup DC) \sqsubseteq c_i. \quad (3)$$

Loop through 2 and 3 until no more inferior or extremal prime cube can be found.
4) If the list U is empty a minimal cover is found; the minimal cover is R. If the list is not empty, covering cycles are present. In such a case, branch to the different possible solutions and retain the one with the lowest cost.

IV. The Procedure in McBOOLE

The procedure implemented in McBOOLE contains basically the same steps as listed above. However, new graph and partitioning techniques are used. The highlights of the new procedure are presented for each of the steps listed.

Generation of Prime Cubes

Efficient ways to generate prime cubes were presented by Reush [15] Morreale [13] and Brayton *et al.* [4]. The recursive generation of prime cubes as defined by Brayton is used here, but is modified to avoid pairwise star-products between cubes in lists. This procedure can be summarized as follows.

The list of cubes representing the function is recursively partitioned along the input variables, up to subfunctions that can be expressed by a single cube. This cube is therefore prime in this sub-space. During partitioning, one of the input variables not already used for partitioning this sub-space is selected. Let $nb0$, $nb1$ and nbx be, respectively, the number of cubes with a 0, 1, and x for an unpartitioned variable. The heuristic used to select the next partitioning variable is

$$\min_j (nbx[j]).$$

When the number of cubes with x is the same for many input variables, the second condition is used:

$$\min_j (\max (nb0[j], nb1[j])).$$

A new node in the partitioning tree is then allocated:

```
structure binary node
  {subtree branch0;
   subtree branch1;
   integer partitioning_variable;
   cube_list listx;
  }
```

All the cubes having an x for the selected variable are split into two cubes having, respectively, a 1 and a 0 instead. All the cubes with a 0 for the partitioning variable are sent to the 0 branch and the cubes with a 1 to the 1 branch. The *listx* is empty for the moment. The recursive procedure to generate prime cubes is performed on the 0 and 1 sub-functions.

At this point, there are two subtrees for this node, each containing all the prime cubes for their respective partitioned sub-space. The next step is to produce the list of prime cubes for the space that includes both those partitioned sub-spaces. Brayton *et al.* [4] do this by pair-wise star-products between all the prime cubes of the two sub-spaces. The new cubes formed are compared together for containment and the covered cubes are deleted. In the present algorithm, the process is roughly the same but some unnecessary trials of star-products between distant cubes are avoided. All the cubes in the 0 subtree are scanned, ($cube_{subtree0}$). For each of them, the 1 subtree is scanned for adjacent cubes. Since cubes in the 1 and 0 subtree differ in the selected variable, they will be adjacent only if all the other variables intersect. At a node in the 1 subtree, only some sub-branches are explored, depending on the value of $cube_{subtree0}$ for the partitioning variable of the node.

If value of variable: 0,   scan 0 subtree and *listx*
1,   scan 1 subtree and *listx*
x,   scan 0 and 1 subtree and *listx*.

This way, only the branches that can contain an adjacent cube are scanned. When an adjacent cube is reached, the newly formed cube is placed in the *listx* of the node at which the sub-spaces are being merged. The new cube is compared with all the cubes already in *listx*. If it is covered by a cube already in the list, the new cube is rejected. If a cube in the list is covered by the new cube, it is deleted from the list. The adjacent cubes, used to form the new cube, are deleted if they are covered by this new cube. When all the star-products are finished, only the prime cubes remain in this space.

Construction of the Covering Graph

A new directed graph is used in McBOOLE to solve the covering problem. When a new cube is formed, it is linked to the cubes that formed it. This way, information about the interaction between the cubes is kept in the graph and makes it possible to solve the covering problem locally. The formation of this new graph is explained below. It is built during the generation of prime cubes. When the termination condition is reached, a single cube remains in a sub-space of the function. The *don't care* part of the cube is extracted and placed in a special list. The *don't care* output bits of the cube are changed to 1 to help the generation of prime cubes. A link ties the *don't care* cube as the ancestor of the single cube in the sub-space. The cube in the sub-space is flagged *basic* and no other cube intersects with it except the *don't care* cube linked to it. All the new cubes are formed during star-product operations:

$$cube_{new} = cube_1 \star cube_2.$$

When a new cube is added to the graph, the combination rules are:

Rule 1: If the new cube covers $cube_1$ and $cube_2$, the ancestors and descendants of $cube_1$ and $cube_2$ are transferred to the new cube. If either $cube_1$ or $cube_2$ was *basic*, the new cube is flagged *basic*.

Rule 2: If $cube_1$ or $cube_2$ is covered, the ancestors and descendants of the covered cube are transferred to the new cube. If the covered cube was *basic* the new cube inherits its status. A new link is created with the new cube, which becomes a descendant of the uncovered cube.

Rule 3: If neither $cube_1$ nor $cube_2$ are covered, the new cube is a descendant of both $cube_1$ and $cube_2$ and is not *basic*.

Some cubes are deleted when they are covered by another cube in a *listx* being formed. When a deletion occurs, a different rule applies.

Rule 4: If the covered cube was *basic*, the cube covering it will inherit its status. The descendants of the covered cube are passed to the cube covering it. The ancestors of the covered cube are simply ignored since they must already be ancestors of the cube that covers it.

The directed graph formed in this way has some interesting properties. Two of those properties will be defined and explained below.

Lemma 1: Only the *basic* cubes may contain a part not shared with any other prime cube and thus may have

$$\forall\, c \in PI, \quad \text{if } c \text{ not basic}, \quad c\,\#\,(PI - c) = \emptyset \qquad (4)$$

Proof: The star-product has the property

$$c_1 \star c_2 = c_3 \text{ implies } c_3 \sqsubseteq \{c_1, c_2\}.$$

When c_1 and c_2 are kept according to Rule 3, c_3 is not *basic* but $c_3\,\#\,\{c_1, c_2\} = \emptyset$, and property (4) is verified. When c_1 is deleted by Rules 1 or 2, it passes its status to c_3. So c_3 is not *basic* only if the cube deleted was not *basic* and was covered by its ancestors. If c_1 is deleted and is not *basic* then

$$c_1\,\#\,\{\text{ancestors of } c_1\} = \emptyset$$

$$\text{so, since } c_3\,\#\,\{c_1, c_2\} = \emptyset$$

$$\text{then } c_3\,\#\,(\{c_2\} \cup \{\text{ancestors of } c_1\}) = \emptyset.$$

The same is true when c_2 is deleted and covered.

When a cube gets deleted because it was covered by another one (Rule 4) it passes its status to the cube covering it.

$$c_1 \sqsubseteq c_2 \quad \text{and} \quad c_1 \text{ is deleted}$$

$$\text{if } c_1 \text{ was not basic } c_1\,\#\,\{\text{ancestors of } c_1\} = \emptyset$$

$$\text{if } c_2 \text{ was not basic } c_2\,\#\,\{\text{ancestors of } c_2\} = \emptyset$$

When c_1 gets deleted, c_2 will not be *basic* only if neither c_1 nor c_2 were *basic*

$$c_2\,\#\,((\{\text{ancestors of } c_2\} - c_1) \cup \{\text{ancestors of } c_1\}) = \emptyset$$

$$\text{since } c_1\,\#\,\{\text{ancestors of } c_1\} = \emptyset.$$

(Q.E.D.)

The graph has a second and very important property.

Lemma 2: All the cubes intersecting with another one are either its descendants or ancestors, or the descendants of its ancestors. Moreover, all the cubes intersecting with a part L of a cube c_a can be found by recursively intersecting c_a only with those direct descendants, and ancestors with their descendants, which intersect with L. This is expressed more formally by the procedure:

let $L = \{c_1, c_2, \cdots, c_i\}$ be a part of c_a

$$\{c_1, c_2, \cdots, c_i\} \sqsubseteq c_a.$$

All cubes c such that $c \sqcap \{c_1, c_2, \cdots, c_i\}$ are found as follows:

```
{
    scan_intersecting_ancestors(c_a)
    scan_intersecting_descendants(c_a)
}

    scan_intersecting_ancestors(cube)
{
    for all direct ancestors c_di
        {if c_di not already scanned and c_di ⊓ {c_1, c_2, ⋯ , c_i}
            {put c_di in the list of cubes intersecting
                scan_intersecting_ancestors(c_di)
                scan_intersecting_descendants(c_di)
            }
        }
}

    scan_intersecting_descendants(cube)
{
    for all direct descendants c_di of cube
        {if c_di not already scanned and c_di ⊓ {c_1, c_2, ⋯ , c_i}
            {put c_di in the list of cubes intersecting
                scan_intersecting_descendants(c_di)
            }
        }
}
```

This property is used extensively in the program to reduce the search time to determine if cubes are extremal or inferior, as defined in (2) and (3).

Proof: When the function is totally partitioned, each partition contains one cube which may be linked to a corresponding *don't care* cube. Since all the partitions are different sub-spaces of the function, all the cubes are disjoint. At that point, *Lemma 2* trivially holds. It must be shown that the property is not altered by the operations that will create or delete cubes in the graph.

When a new cube is formed, if neither c_1 or c_2 are covered, Rule 3 applies:

$$c_1 \star c_2 = c_3, \quad c_3 \text{ is a descendant of } c_1 \text{ and } c_2$$

if $c_3 \sqcap L$ a list, then c_1 or $c_2 \sqcap L$ since $c_3 \sqsubseteq \{c_1, c_2\}$.

When, say, c_1 is scanned for a list L, c_3 will also get scanned if it intersects L, since it is a direct descendant of c_1. When c_3 is scanned for L, c_1 will get scanned if c_1

$\sqcap L$. The same thing could, by symmetry, be said for c_2. Also, since it is assumed that the property was verified before c_3 was added to the graph, all the cubes intersecting with L will be scanned, including c_3, because c_1 and c_2 are direct ancestors of c_3.

When c_1 is covered and deleted (Rule 2), its ancestors and descendants are passed to c_3. This way, c_3 simply takes the place of c_1 in the graph, and nothing is changed regarding the cubes intersecting with c_1. The cube c_2 intersects with c_3 but c_2 is kept and linked as a direct ancestor of c_3, and this case is equivalent to Rule 3. The same applies, by symmetry, when c_2 is covered and c_1 is kept.

When both c_1 and c_2 are covered and deleted (Rule 1),

---

c_3 takes the place of both in the graph, and nothing is affected.

When a cube is deleted because it is covered by another one (Rule 4), its descendants are passed to the cube covering it. Suppose c_1 gets covered by c_2 and is deleted. Before c_1 was deleted, all the cubes intersecting with L could be found when $c_1 \sqcap L$.

$$\text{since } c_1 \sqsubseteq c_2$$

$$\text{if } c_1 \sqcap L \text{ then } c_2 \sqcap L.$$

When c_1 is deleted, and L is being scanned, the ancestors of c_1 will get scanned anyway since the property holds for c_2 and $c_1 \sqsubseteq c_2$. Also c_2 will get scanned since it intersects with L. When c_2 is scanned, the descendants of c_1 will also get scanned since they were passed to c_2 when c_1 was deleted. Q.E.D.

It has been shown that the formation of new cubes and

the deletion of cubes does not affect the properties stated in Lemma 1 and Lemma 2, provided the rules are used to modify the graph when a cube is added or removed. So at the end of the generation of the prime cubes, a graph that links together related cubes is obtained. Also some of the cubes have the special status *basic*. The reader should note that the generation of prime cubes as well as all the definitions apply to multiple output cubes are presented in [7].

Extraction of a Minimal Cover

All the prime cubes, as well as the list of *don't care* cubes, have now been obtained. A subset of the prime cubes will form the minimal cover. The status of the cubes can be one of three: *undecided, unretained,* or *retained.* The cubes that will be *retained* will compose the final solution. All the cubes start with the status *undecided.* Associated with each cube is an uncovered part defined as

$$\text{uncov}(cube) = cube \# (\{\text{retained cubes}\} \cup DC).$$

At the beginning, the uncovered part of a cube is the cube itself. The space of the function is divided in two parts: the part covered by the *retained* cubes and that covered only by the *undecided* cubes. The solution is reached when the whole function is covered by the *retained* cubes.

When the cover extraction procedure is started, no prime cube is inferior (property (3)) by definition (1) of a prime cube, since all the prime cubes are *undecided.* Also, since at the beginning all cubes are *undecided,* only the essential prime cubes will be extremal. An essential prime cube is defined by

$$c_i \in \{\text{prime cubes}\}, \quad c_i \# (\{\text{prime cubes}\} - c_i) \neq \emptyset \quad (5)$$

To begin with, the *don't care* cubes are sharped from the uncovered part of the intersecting cubes:

$$\text{uncov}(cube) = \text{uncov}(cube) \, \# DC.$$

All the cubes that have their uncovered part updated get a special status *affected retained.*

Next all the essential prime cubes are identified and retained. Only the *basic* cubes can satisfy relation (4) from *Lemma 1,* and be essential prime. For each of them L is computed:

$$L = c_i \# \{\text{direct ancestors and descendants of } c_i\}$$

If $L = \emptyset$, c_i is not an essential prime cube. When $L \neq \emptyset$, since no direct ancestors or descendants intersect with the remaining part, no other cube can intersect with it from *Lemma 2.* The cube is then essential and its status is set to *retained.* All the cubes intersecting with c_i get c_i sharped from their uncovered part. The cubes that have their uncovered part updated are flagged *affected retained.*

The next step in the procedure is to find the inferior cubes (Definition (3)). A cube c is inferior or equal to a cube c_j if:

$$c, \exists c_j, \quad c \neq c_j$$

$$c_j \in \{\text{undecided cubes}\}, \quad \text{uncov}(c) \sqsubseteq c_j.$$

Lemma 3: A cube can become inferior only when its uncovered part is modified.

Proof: Initially, the uncovered part of a cube was the cube itself, and no prime cube was inferior by definition (1). So each cube satisfies the relations:

$$\text{uncov}(c) \neq \emptyset$$

$$\forall c_j \in (U - c), \quad \text{uncov}(c) \not\sqsubseteq c_j. \quad (6)$$

Since no new cube can be added to the list U, the only way a cube can become inferior is when its uncovered part is modified. Furthermore, since all the cubes having their uncovered part updated are flagged *affected retained,* only those need to be examined to find all the inferior cubes.
Q.E.D.

All the *affected retained* cubes are next examined to identify the inferior cubes (definition (3)). When a cube is inferior, it is removed from the list U. All the cubes intersecting with the uncovered part of the *unretained* inferior cube will get the special status *affected unretained.* If the cube is not inferior, its status is cleared and it cannot become inferior and will not be examined again until it gets *affected* again.

Some extremal cubes now have to be found. Initially, the essential prime cubes were identified and retained. None of the remaining cubes were extremals.

Lemma 4: A cube can become extremal only when an *unretained* cube intersects with its uncovered part.

Proof: Relation (2) which defines extremals, can be expressed in a different but strictly equivalent form:

$$\text{uncov}(c) \# (\{\text{prime cubes}\} - c - \{\text{unretained}\}) \neq \emptyset. \quad (7)$$

It was shown (*Lemma 1*) that only the basic cubes could be extremal at the beginning, since no cubes were *unretained.* Now all the basic cubes have been examined and the extremals have been identified and *retained.* None of the remaining cubes can satisfy (7) since no cubes were *unretained.* Also, this relation can be true for a cube only when another cube intersecting with its uncovered part is *unretained* and removed from U. All the cubes intersecting with the uncovered part of an *unretained* cube were set to *affected unretained* and, therefore, only those cubes can become extremal.
Q.E.D.

All the *affected unretained* cubes are scanned next to identify the extremals. The extremal cubes are removed from U and retained in R. The cube *retained* is sharped from the uncovered part of the intersecting cubes. When a cube gets its uncovered part modified, it gets the status *affected retained.* The cubes which are not extremal have their status cleared and cannot become extremal and will not be examined until they get *affected* again.

This notion of *affected* cubes appears to be new and has been introduced to avoid unnecessary processing of cubes that could not be extremal or inferior. In this way, only cubes which are very likely to be extremals or inferior are examined, and a lot of computation time is saved. As the minimization proceeds, decisions on cubes are made, new ones get *affected* until, for all the remaining cubes, the following relations apply:

$$\text{uncov}(c) \, \# \, (\{\text{undecided cubes}\} - c) = \emptyset$$

$$\forall \, c_j \in \{\text{undecided cubes}\}, \quad c \neq c_j, \quad \text{uncov}(c) \not\sqsubseteq c_j.$$

If U is empty at that point, a minimal solution has been reached. The minimal cover is in R. However, if U is not empty, conditions defining covering cycles have been encountered.

Partitioning and Branching Cycles

At this point, the only way to guarantee a minimal cover is to perform branching. An *undecided* cube is picked, retained, and the uncovered part of cubes that will get *affected retained* is updated. The solution that follows from this choice is computed. Next, the initial cycle condition is restored, the same *undecided* cube is *unretained* and the *affected unretained* cubes are appropriately flagged. This alternate solution is computed and, finally, both solutions are compared. The solution with the lowest cost is kept. Note that when these solutions are computed, cubes get *affected* and the procedure explained before applies recursively. In particular, nested cycles can be encountered and many branches might have to be computed and compared.

In the present algorithm, a new partitioning scheme is used that can greatly reduce the number of branches. Partitions are branched and solved independently. When nested cycles are encountered in a partition, they are also recursively partitioned and branched. When the undecided list with cycles is obtained, one cube is picked. All the cubes related with this cube and to each other are put in a partition, which is then solved. From the undecided cubes left, another partition is built and solved. This continues until the undecided list is empty. The procedure to build a partition is:

```
{
  select a cube c
  put it in the partition
  scan_partition(c)
}

  scan_partition(cube)
{
  for all c_d direct ancestors and descendants of cube
    {if c_d not already scanned and c_d ⊓ uncov(cube)
      {put c_d in the partition
       scan_partition(c_d)
      }
    }
}
```

Let c_{pi} be the cubes in the partition and c_{nj} the cubes that are not. The following relation then holds

$$\forall \, (c_{pi}, c_{nj}), \quad \text{uncov}(c_{pi}) \not\sqcap c_{nj} \quad \text{or} \quad \text{uncov}(c_{nj}) \not\sqcap c_{pi}.$$

$$(8)$$

Lemma 5: The cubes, satisfying relation (8) that are not in the solved partition are not affected by the decisions taken for the cubes in the partition. Therefore the covering problem in a partition can be solved independently from the remaining cubes.

Proof: The only cubes that are affected by a decision on a cube c are, from *Lemmas 3 and 4*, $c_i \sqcap \text{uncov}(c)$. In that case, when a decision is taken on any node in the partition, no cube outside the partition (8) is affected, and the two sets of cubes are, therefore, completely independent.

Q.E.D.

It is very important to differentiate independent cycles from nested cycles since, in the first case, the number of branches grows linearly and, in the second, exponentially with the number of cycles encountered. In a typical case examined, the branching depth reached 12 without partitioning and only 4 with this new partitioning scheme. Both ways obviously lead to the same solution, but the partitioning saves a lot of CPU time when independent cycles are encountered.

V. Results

The algorithm described above was programmed in C language under UNIX 4.2BSD, and the results presented were obtained on a VAX11/750. The program is highly portable and should run easily on most computers. Many different logic functions, used in industrial PLA's and provided with the program ESPRESSO IIC [6], were minimized. Table I summarizes the results and compares them to those obtained with ESPRESSO IIC on the same computer. For all these examples Table I shows the number of input and output variables, the number of prime cubes and the number of essential prime cubes. The table also shows the number of cubes for the unminimized function, minimized using McBOOLE and, finally, minimized using ESPRESSO IIC. The CPU time and the memory consumed by each program are also included. The examples in the table are sorted by increasing $time_{\text{McBOOLE}}/time_{\text{ESPRESSO}}$. Thus the examples at the top of the list were functions that could be minimized much faster by McBOOLE than by ESPRESSO IIC, whereas those at the end of the table were relatively difficult for McBOOLE.

In all cases, McBOOLE produced a solution with a lower or equal number of cubes, since it insures a minimal cover. ESPRESSO IIC, on the other hand, uses only heuristics but is quite efficient and, in almost all the cases examined, reached the minimal solution or very near to it.

A careful study of these results helps to identify the factors that affect most the computational requirements of the two algorithms. The major factors identified are listed below.

Storage

The storage requirements of ESPRESSO IIC seem to be most strongly correlated to:

The size of the function (# cubes × the number of variables);

TABLE I
COMPARISON OF McBOOLE (mc.) AND EXPRESSO IIC (esp.)

name	#input variables	#output variables	#cubes in	#prime cubes	#essential cubes	#cubes out mc.	#cubes out esp.	memory in K mc.	memory in K esp.	CPU seconds mc.	CPU seconds esp.
xor12	12	1	2048	2048	2018	2048	2048	431	6112	10.6	4167.2
xor10	10	1	512	512	512	512	512	190	998	8.4	103.0
pla2x	50	69	183	953	145	178	178	571	2590	141.3	2431.3
dk14mv	15	12	107	56	56	56	56	111	436	1.8	29.3
dk15mv	12	9	68	32	32	32	32	100	320	1.0	11.8
dk17mv	12	11	68	32	32	32	32	100	318	1.0	11.2
simple	9	6	38	20	20	20	20	94	268	0.6	4.7
dk512mv	17	18	138	30	30	30	30	110	348	1.8	13.9
dk27mv	9	9	38	11	14	14	14	90	232	0.5	3.3
col4	14	1	47	14	14	14	14	92	284	0.6	3.6
pi2	4	3	14	19	7	12	12	104	246	0.6	3.0
dc1	4	7	15	22	3	9	9	104	262	0.7	3.1
mult2	4	4	12	12	2	7	7	104	242	0.5	2.2
pi	5	5	10	15	5	8	8	104	244	0.6	2.6
add2	4	3	17	17	7	11	11	104	256	0.6	2.5
example1	3	4	4	5	3	4	4	104	238	0.4	1.5
add4	8	5	397	397	35	75	75	220	570	27.0	83.0
dist	8	5	256	401	23	120	121	272	666	48.3	129.9
in0	15	11	138	706	60	107	107	304	584	19.3	121.6
dc2	8	7	58	173	18	39	10	132	324	6.5	15.9
adr4	8	5	256	397	35	75	75	206	618	27.5	59.9
bigpla	10	10	93	427	20	85	85	238	186	11.2	80.8
gary	15	11	214	706	60	107	107	310	592	61.4	106.6
f51m	8	8	256	561	13	76	76	278	820	63.1	102.3
in1	16	17	110	928	54	104	101	328	812	118.1	170.0
apla	10	12	134	201	0	25	26	238	424	31.3	27.1
in2	19	10	137	666	85	134	137	514	558	184.3	114.3
dk17	10	11	93	111	0	18	18	250	318	31.1	18.1
ryy6	16	1	112	112	112	112	112	288	582	150.6	79.3
dk27	9	9	52	82	0	10	10	246	326	33.6	15.6
alu2	10	8	91	134	36	68	68	416	431	189.4	73.7
mult4 *	8	8	606	606	12	124	133	516	670	859.1	305.7
alu3	10	8	72	540	27	64	66	536	418	296.8	61.3
clpl	11	5	20	143	20	20	20	224	256	28.0	5.3
pope*	6	48	64	593	12	62	63	642	806	5239.1	393.3
sqr6 **	6	12	280	205	3	19	50	290	121	972.7	54.0
alu1	12	8	19	780	19	19	19	656	232	531.3	1.4

*branching abandoned after 6 nested cycles
**branching abandoned after 8 nested cycles

The size of the complement of the function.
For McBOOLE the storage requirements are directly correlated to the number of prime cubes and the number of variables in the function.

CPU Time

The CPU time of ESPRESSO is mostly related to:

The number of cubes in the function which are not essential prime cubes, since the latter are identified early in the procedure and need far less computation;
The number of variables in the function;
The size of the complement of the function. This will affect the expansion of cubes, since these cubes are expanded until a cube in the complement intersects with them.

For McBOOLE the CPU time is related to:

The number of prime cubes;
The number of nested cycles in the function.

In some difficult cases, the number of nested cycles can exceed a limit fixed by the user, whereupon the branching in McBOOLE is abandoned. In such a case the user is warned that the minimal solution is not guaranteed. Three such cases are presented in the table and are identified. It is interesting to see that, at least in these three cases, even though a minimal cover could not be insured by Mc-BOOLE, the solution obtained was superior to the solution generated by ESPRESSO; the difference between the solutions obtained by the two programs is particularly interesting in the case of *mult*4.

Because of their different requirements, the two minimizers take very different amounts of CPU time depending on the function. It is apparent that when the number

of prime cubes is not much larger than the initial number of cubes, McBOOLE is faster. For functions where the number of cubes is large but they are not too closely related, the number of prime cubes is not too big. The graph and partitioning techniques are very efficient for these loosely coupled cubes. A very good example to illustrate this is the 12 input exclusive or (xor12) function. For that case, McBOOLE performed the minimization more than 100 times faster than ESPRESSO IIC.

On the other hand, for functions having a high number of prime cubes, McBOOLE is severely penalized because it generates all the prime cubes. The example *alu*1 illustrates this very well. Since one cannot know in advance the number of prime cubes, and it is not always easy to see if cubes are loosely coupled, the following rules can be used to select algorithms. It has been observed that when the number of either input or output variables is large the problem gets more difficult; a problem with 5 input and 50 output variables will be, in general, much more difficult than a problem with 10 input and 10 output variables. Because of that, we will refer to functions with, for example, up to 10 input and 10 output variables as simply functions with up to 10 variables.

For up to 10 variables the number of prime cubes is usually reasonable and McBOOLE is generally faster than ESPRESSO IIC.

For up to 20 variables, the CPU time of McBOOLE is usually in the same range as ESPRESSO IIC for most cases where the number of prime cubes does not grow too much. McBOOLE is, therefore, still very attractive for these cases, since it always produces a minimal cover. There will always be, however, some difficult functions with less than 20 variables for which an exact solution is impractical.

Bigger functions must be examined carefully to determine if McBOOLE is suitable. Some functions like *pla*2*x* in the table were easily minimized although they contained more than 50 input and 50 output variables.

The results obtained in the comparisons of McBOOLE with ESPRESSO IIC and other minimizers, lead to the following conclusions.

The previously reported limit of 10 input variables [6], [10] for producing a minimal cover, has been improved to around 20 input variables for most typical industrial PLA's. The features of McBOOLE that lead to this improvement are:

a) A more efficient generation of prime cubes which avoids pair-wise star products.
b) The covering graph gives all the intersecting cubes more rapidly.
c) Because the *affected* cubes are flagged, many cubes are not processed needlessly.
d) Finally, the cycles are partitioned and branched independently.

The results obtained are also useful because they provide an evaluation of ESPRESSO IIC. It has been verified that it gives generally very near to minimal solutions for the medium to large functions for which the minimal cover could be obtained with McBOOLE. Therefore, one can be confident that ESPRESSO IIC most probably also gives good results for very large functions that currently cannot be handled by McBOOLE.

VI. Further Improvements

The very high number of prime cubes was the limitation in most of the examples that McBOOLE could not handle. There were also some other cases where the number of nested cycles was unmanageable and the branching had to be abandoned. The heuristic minimizers ESPRESSO IIC and MINI II can handle much larger functions because they do not generate all the prime cubes and use some very efficient logic operators like Tautology [4]. Some tests exist in ESPRESSO IIC and MINI II to identify the essential prime cubes without generating all the prime cubes.

A procedure has been developed by the authors to form and identify the other extremal cubes, by only generating and examining a small subset of the prime cubes, locally around a cube being examined. This new procedure has very low memory requirements since only a small subset of the prime cubes is stored at any time. It has also the interesting property that the initial list of cubes is iteratively improved and the procedure can be stopped at any point yielding a good solution.

However, the branching cannot easily be implemented using this approach so the minimal solution is found only when no covering cycles are present. Also, even though the memory requirements are improved, it appears that many prime cubes have to be generated locally more than once, since they are not stored. The CPU time required by the new procedure is not affected much by the fact that not all the prime cubes are generated.

The goal of McBOOLE was to minimize the number of cubes. It is also desirable to reduce the number of literals, or the number of transistors, in a PLA. Reducing the number of transistors can improve the foldability of a PLA. Also it can improve the speed of a PLA if the removed transistors are in the critical path. Therefore, the minimization of the number of literals is a useful secondary goal; for a given number of cubes, try to minimize the number of literals. One way to improve McBOOLE would be to take its output and pass it to a very fast minimizer, like in [7], that reduces the number of literals. ESPRESSO includes some steps at the end to reduce the number of literals, and, for a given number of cubes, very often produces a solution which, compared with McBOOLE's, has fewer literals.

Appendix I
McBOOLE Procedure for Logic Minimization

```
main( )
    {
        read cubes
        generate_prime_cubes_by_recursive_partitioning( )
        Set the status of all cubes to undecided
        For all cubes set uncov(c) = c
        For all the don't care cubes retain(cube)
        For all basic cubes, if (c # {direct ancestors, descendants} ≠ ∅) retain(c)
        Solve(remaining undecided cubes)
    }

Solve(list of cubes)
    {
        Until the affected queue is empty
        {cube = get next cube from queue
            if the cube is affected retained
                {if (uncov(cube) = ∅) discard the cube
                if (uncov(cube) ⊑ c_i, c_i undecided) unretain_inferior(cube)
                else clear the cube status
                }
            if the cube is affected unretain
                {if (uncov(cube) # {undecided cubes} ≠ ∅) retain(cube)
                else clear the cube status
                }
        }
        Until undecided list is not empty
        {Take an undecided cube
            put in partition all related cubes: scan_partition(cube)
            branching_cube = select_branching_cube(partition)
```

Store the partition
retain (*branching_cube*)
solve (*partition*)
store the solution and restore the original partition
unretain_inferior (*branching_cube*)
solve (*partiton*)
compare the two solutions and keep the best one
}
return ( )
}

retain (*cube*)
{For all $c_i \in U$, *if* $c_i \sqcap uncov$ (*cube*)
{If c_i not already in queue, put it on the queue
$uncov(c_i) = uncov(c_i) \# cube$ its status is set to *affected retained*
}
If *cube* is a *don't care* cube, remove it from U
else transfer *cube* from U to R
}

unretain_inferior(cube)
{For all $c_i \in U$, *if* $c_i \sqcap uncov$ (*cube*)
{If c_i not already in queue, put it in the queue.
The status of c_i is set to *affected unretain*
}
remove *cube* from the *undecided* list
}

ACKNOWLEDGMENT

The authors are grateful to Ajoy Bose and Michael Thong for inviting the first author to AT&T Bell Labs where he was able to compare McBOOLE to several logic minimizers including ESPRESSO IIC. While there he had stimulating discussions with Jean Dussault and Prathima Agrawal. In particular, Jean Dussault also helped in the evaluation of McBOOLE and ESPRESSO IIC.

REFERENCES

[1] Z. Arevalo and J. G. Bredeson, "A method to simplify a Boolean function in a near minimal sum-of-products for programmable logic arrays," *IEEE Trans. Computers*, vol. C-27, p. 1028, Nov. 1978.

[2] M. Bartholomeus and H. de Man, "PRESTOL-II: Yet another logic minimizer for programmed logic arrays," in *Proc. 1985 Int. Symp. Circuits and Systems*, Kyoto, Japan, June 1985.

[3] N. Biswas, "Computer aided minimization procedure for Boolean functions," in *Proc. 21st Design Automation Conf.*, p. 699, June 1984.

[4] R. K. Brayton, J. D. Cohen, G. D. Hachtel, B. M. Trager, and D. Y. Y. Yun, "Fast recursive Boolean function manipulation," in *Proc. 1982 Int. Symp. on Circuit and Systems*, p. 58, May 1982.

[5] R. K. Brayton, G. D. Hachtel, C. T. McMullen, and A. Sangiovanni-Vincentelli, "ESPRESSO II a new logic minimizer for PLA," in *IEEE Proc. Custom Integrated Circuits Conf.*, p. 370, May 1984.

[6] R. K. Brayton, G. D. Hachtel, C. T. McMullen, and A. Sangiovanni-Vincentelli, *Logic Minimization Algorithms for VLSI Synthesis*, Kluwer Academic, Boston, 1984.

[7] D. W. Brown, "A state machine synthesizer SMS," in *Proc. 18th Design Automation Conf.*, p. 301, June 1981.

[8] L. Dietmeyer, *Logic Design of Digital Systems*. Boston, MA: Allyn and Bacon, Mar. 1979.

[9] S. J. Hong, R. G. Cain, and D. L. Ostapko, "MINI: A heuristic approach for logic minimization," *IBM J. Res. Develop.*, pp. 443–458, Sept. 1974.

[10] K. Ishikawa, T. Sasao, and H. Terada, "A minimization algorithm for logical expressions and its bounds of application," *Trans. IECE Japan*, vol. J-65D, p. 797, June 1982.

[11] E. J. McCluskey, "Minimization of Boolean functions," *Bell Syst. Tech. J.*, vol. 35, p. 1417, Apr. 1956.

[12] M. J. Meyer, P. Agrawal, R. G. Pfister, "A VLSI FSM design system," in *Proc. 21st Design Automation Conf.*, p. 434, June 1984.

[13] E. Morreale, "Recursive operator for prime implicant and irredundant normal form determination," *IEEE Trans. Computers*, vol. C-19, p. 504, June 1970.

[14] W. V. Quine, "A way to simplify truth functions," *Amer. Math. Monthly*, vol. 62, p. 627, Nov. 1955.

[15] B. Reush, "Generation of prime implicants from subfunctions and a unifying approach to the covering problem," *IEEE Trans. Computers*, vol. C-24, p. 924, Sept. 1975.

[16] T. Rhyme, P. S. Noe, M. H. McKinney, and U. W. Pooch, "A new technique for the fast minimization of switching functions," *IEEE Trans. Computers*, vol. C-26, p. 757, Aug. 1977.

[17] J. P. Roth, R. M. Karp, "Minimization over Boolean graphs," *IBM J.*, vol. 6, no. 2, p. 227, Apr. 1962.

[18] T. Sasao, "Input variable assignment and output phase optimization of PLAs," *IEEE Trans. Computers*, vol. C-33, p. 879, Oct. 1984.

[19] A. Svoboda, "The concept of term exclusiveness and its effect on the theory of Boolean functions," *J. ACM*, vol. 22, no. 3, p. 425, July 1975.

[20] P. Tison, "Generalization of consensus theory and application to the minimization of Boolean functions," *IEEE Trans. Electron. Computers*, vol. EC-16, p. 446, Aug. 1967.

[21] M. R. Dagenais, V. K. Agarwal, N. C. Rumin, "The McBOOLE logic minimizer," in *Proc. 22nd Design Automation Conf.*, p. 667, Las Vegas, NV, June 1985.

Symbolic Design of Combinational and Sequential Logic Circuits Implemented by Two-Level Logic Macros

GIOVANNI DE MICHELI, MEMBER, IEEE

Abstract—This paper presents a method for the optimal synthesis of combinational and sequential circuits implemented by two-level logic macros, such as programmable logic arrays. Optimization consists of finding representations of switching functions corresponding to minimal-area implementations. The design of optimization is based on two steps: symbolic minimization and constrained encoding. Symbolic minimization yields an encoding-independent *sum of products* representation of a switching function which is minimal in the number of product terms. The minimal symbolic representation is then encoded into a compatible Boolean representation. The algorithms for symbolic minimization and the related encoding problems are described. The computer implementation and the experimental results are then presented.

I. INTRODUCTION

THE AUTOMATED synthesis of regular modules for very large scale integrated (VLSI) circuit design decreases design time and ensures functional correctness. In order to compete with manual designs, computer-aided synthesis techniques must include optimization procedures which attempt to minimize both silicon area and circuit switching times.

We present here a method for the optimal synthesis of digital modules implementing combinational and/or sequential switching functions. Modules are implemented by two-level logic macros, such as programmable logic arrays (PLA's) and synchronous registers. Optimization consists of finding representations of switching functions at the logic level that minimize the silicon area taken by their physical implementation (referred to as *cost* in short), without considering the interconnection area. We assume that the digital modules to be implemented can be described by tables. In general, tabular descriptions can be obtained in a straightforward way from structural-level systems descriptions in Hardware Description Languages (HDL), as in the case of the Yorktown Silicon Compiler [3].

In the standard approach to synthesis [10], Boolean representations of switching functions are obtained from the structural description by representing each mnemonic entry in a table (or each variable in a HDL program) by Boolean variables. The optimization of logic functions (and, in particular, two-level logic minimization) is performed on the Boolean representation. The result of logic optimization is heavily dependent on the representation of the variables. As an example, the complexity (in particular, the minimal cardinality of a two-level implementation) of the combinational component of a finite-state machine depends on the assignment of Boolean variables to he internal states [9].

The **symbolic design** methodology presented here avoids the dependence on the variable representation in the optimization process and consists of two steps: 1) determine an optimal representation of a switching function independently on the encoding of its inputs and outputs; 2) encode the inputs and outputs so that they are compatible with the oꞇtimal representation. This technique can be applied to sꞁlve the following problems of logic design:

P1) find an encoding of the inputs (or some inputs) of a combinational circuit that minimizes its cost;

P2) find an encoding of the outputs (or some outputs) of a combinational circuit that minimizes its cost;

P3) find an encoding of both the inputs and the outputs (or some inputs and some outputs) of a combinational circuit that minimizes its cost;

P4) find an encoding of both the inputs and the outputs (or some inputs and some outputs) of a combinational circuit that minimizes its cost and such that the encoding of the inputs is the same as the encoding of the outputs (or the encoding of some inputs is the same as the encoding of some outputs).

Finding an optimal state assignment of a sequential circuit is equivalent to solving problem P4, when the sequential circuit is implemented by feeding back (possibly through registers) some outputs of a combinational circuit to its inputs. Similarly, finding the encodings of the signals connecting two (or more) combinational circuits, that minimize the total cost, can be reduced to problem P4. The author presented in [4] and [5] an approximation to the solution of the state assignment problem, in which the

Manuscript received January 29, 1986; revised May 19, 1986. This research is a part of the project "Research on topological design tools for structured logic arrays," sponsored by the National Science Foundation under Contract ECS-8121446.

The author is with the IBM Thomas J. Watson Research Center, Yorktown Heights, NY 10598.

IEEE Log Number 8609826.

Reprinted from *IEEE Trans. CAD of Int. Circ. Syst.*, vol. CAD-5, no. 4, pp. 597–616, Oct. 1986.

cost was minimized with regard only to the encoding of the inputs. In particular, the technique presented in [4] and [5] solved only problem P1. Problem P2 was attacked by Nichols [13], but the algorithm he presented could deal only with small-scale circuits. We refer the reader to [5] for an extended set of references and a critical survey of most of the previous techniques for state assignment.

Though the symbolic design methodology is fairly general, we restrict our attention here to two-level *sum of products* implementations. Since the area of the physical implementation has a complex functional dependence on the function representation (even by using PLA implementations [5], we consider a simplified optimization technique that leads to quasi-minimal areas. In particular, we attempt to find first *a sum of products* representation that is minimal in the number of products, and then a representation of the input/output that is minimal in the number of Boolean variables.

The difficulty in solving problems P2–P4 is related to finding a minimal two-level representation of a switching function independently of the encoding of both inputs and outputs. We introduce here a technique called **symbolic minimization**. Symbolic minimization consists of determining a minimal encoding-independent two-level *sum-of-products* representation of a switching function. It is minimal in number of product terms and independent of the encoding of all (or part of) the inputs and outputs [6]. The minimal symbolic representation is an intermediate step towards the determination of a corresponding Boolean representation. For this reason, three encoding problems are introduced to transform the minimal symbolic cover into an equivalent Boolean representation.

Section II contains a general overview of the symbolic design methodology and is an informal introduction to the problem. Then, in Section III, we present in detail the properties of the symbolic representation and an algorithm for symbolic minimization. The three new encoding problems are introduced in Section IV, as well as an algorithm for constrained encoding. The computer implementation of the symbolic minimization and the encoding algorithms is also presented along with experimental results.

II. OVERVIEW

In this section, we present an informal overview of symbolic design. The methodology is introduced by elaborating on an example. We consider first a combinational circuit (Fig. 1) and we use symbolic design to find a representation of its inputs and outputs and a corresponding two-level implementation that minimize its cost. This is equivalent to solving problem P3 of Section I. Note that problem P1 or P2 can be derived from problem P3 by considering only the circuit inputs or ouputs.

Example 1: The following truth table specifies a combinational circuit; in particular, an instruction decoder. There are three fields: the first is related to the addressing mode, the second to the operation code, and the third one to the corresponding control signal. The circuit has two

Fig. 1.

inputs and one output. Each row specifies a symbolic output for any given combination of symbolic inputs

INDEX	AND	CNTA
INDEX	OR	CNTA
INDEX	JMP	CNTA
INDEX	ADD	CNTA
DIR	AND	CNTB
DIR	OR	CNTB
DIR	JMP	CNTC
DIR	ADD	CNTC
IND	AND	CNTB
IND	OR	CNTD
IND	JMP	CNTD
IND	ADD	CNTC

In the standard approach to synthesis, each word (mnemonic string) in the table would be encoded by a string of binary symbols (i.e., 1's and 0's). Then, the encoded table would be minimized by a logic minimizer. In symbolic logic design, the table is minimized first (i.e., a table consisting of a minimal number of rows is computed) and then the table is encoded.

A first approach to symbolic minimization can be achieved by grouping the set of inputs that correspond to each output symbol. This process of reducing the size of the table is called here **disjoint minimization** because the table is considered as a set of independent subtables corresponding to each output symbol. Remarkably, disjoint minimization can be achieved by using techniques of multiple-valued logic minimization [2], [5], as shown in Section III.

Example 2: From the table of Example 1, we can see that the addressing mode INDEX and any operation codes AND OR ADD JMP correspond to the control CNTA. Similarly either one of the following conditions

addressing mode DIR and operation codes AND or OR
addressing mode IND and operation code AND

correspond to control CNTB. The entire table can be expressed as a set of conditions

INDEX	AND OR ADD JMP	CNTA
DIR	AND OR	CNTB
IND	AND	CNTB
IND	OR JMP	CNTD
DIR IND	ADD	CNTC
DIR	JMP	CNTC

Note that this table is more compact than the previous one, because it requires only six rows instead of twelve.

The problem now is to find a Boolean representation of the symbols, corresponding to a Boolean cover representation of the function, with as many rows as the compacted table. While the encoding problem will be presented in detail in Section IV, we show here by an example the consequence of the choice of a particular encoding.

Example 3: Consider this particular Boolean encoding of the words

$$\begin{array}{lll}
\text{INDEX} = 00 & \text{AND} = 00 & \text{CNTA} = 11 \\
\text{DIR} = 01 & \text{OR} = 01 & \text{CNTB} = 01 \\
\text{IND} = 11 & \text{ADD} = 10 & \text{CNTC} = 10 \\
& \text{JMP} = 11 & \text{CNTD} = 00
\end{array}$$

Then the function can be represented by a Boolean cover as

$$\begin{array}{lll}
00 & ** & 11 \\
01 & 0* & 01 \\
11 & 00 & 01 \\
11 & *1 & 00 \\
*1 & 10 & 10 \\
01 & 11 & 10
\end{array}$$

where a *don't care* condition on a binary input variable is represented by *. Note that the fourth Boolean implicant can be deleted because its output part is 00. Moreover, note that by deleting this implicant this cover is minimum, i.e., there exist no Boolean covers corresponding to this encoding with fewer than five implicants. (This can be proven experimentally by running an exact minimization algorithm [14], like that implemented by program ESPRESSO-II with the "exact" flag [18].)

We question now to what extent symbolic design guarantees the minimality[1] of the Boolean cover, obtained by replacing the words by their corresponding binary encoding.

Example 4: Consider this other Boolean encoding of the words, in which we changed only the encoding of the output symbols

$$\begin{array}{lll}
\text{INDEX} = 00 & \text{AND} = 11 & \text{CNTA} = 00 \\
\text{DIR} = 01 & \text{OR} = 01 & \text{CNTB} = 01 \\
\text{IND} = 11 & \text{ADD} = 10 & \text{CNTC} = 10 \\
& \text{JMP} = 11 & \text{CNTD} = 11
\end{array}$$

Then the function can be represented by a Boolean cover as

$$\begin{array}{lll}
00 & ** & 00 \\
01 & 0* & 01 \\
11 & 00 & 01 \\
11 & *1 & 11 \\
*1 & 10 & 10 \\
01 & 11 & 10
\end{array}$$

[1]The optimality of a Boolean cover is measured by its cardinality, i.e., by the number of its implicants. A Boolean cover of a function is **minimum** if there exists no cover of that function having a smaller cardinality. A Boolean cover of a function is **minimal** if its cardinality is minimum with regard to some local criterion. Usually, a Boolean cover of a function is said to be minimal if no proper subset is a cover of the same function [2].

Note that the first Boolean implicant can be deleted because its output part is 00. However, note that this cover is not minimum: there exist now a minimum Boolean cover corresponding to this encoding with three implicants and that can be computed from the above one by a standard minimization technique, namely

$$\begin{array}{lll}
1 & 0 & 01 \\
1 & 1 & 10 \\
11 & *1 & 11
\end{array}$$

Note that the first cover has two pairs of Boolean implicants with 01 and 10 in the third field and that are merged into two single implicants in the minimum cover. This is possible because the third implicant of the minimum cover has 11 in the third field, and 11 covers 01 and 10.

The reason for this additional reduction is in the covering relations among the encoded output symbols. Note that when we optimized the symbolic table by disjoint minimization, our goal was only to group the input symbols corresponding to each output symbol independently. The relations among the output symbols were neglected. For this reason, it is important to exploit the relations among the output symbols at the symbolic level. **Symbolic minimization**, formally defined in Section III, is a technique that determines an optimal ordering of the output symbols. This ordering is related to the covering relations among the binary encodings of the output symbols, and is responsible for the additional reduction of the table size, as shown by Example 4.

Example 5: Consider the following table:

INDEX	AND OR ADD JMP	CNTA
DIR IND	AND OR	CNTB
DIR IND	ADD JMP	CNTC
IND	JMP OR	CNTD

Here, an ordering relation is assumed that allows control CNTD to override control CNTB and CNTC when both are specified. The table, together with this ordering relation, is an equivalent representation of the function specified by Example 1. It is an example of the result of a symbolic minimization. Note that this table can be transformed into the Boolean cover of Example 4 by replacing each symbol by its encoding. Moreover, the encoding of CNTD covers bit-wise the encoding of CNTB and CNTC and allows CNTD to override CNTB and CNTC. Note that the first implicant can be deleted by assuming that CNTA is the default output and that the encoding of CNTA is 00, which is covered bit-wise by the encoding of all other outputs.

Once a minimal table has been found, the encoding of the words into Boolean variables is driven by the grouping of the input symbols (group of symbols appear on the same row of a minimal table) and the ordering of the output symbols generated by symbolic minimization. Note that disjoint minimization deals with each output symbol independently, and therefore does not provide information for an encoding of the output symbols that optimize the table size. Therefore, disjoint minimization can be used only to solve problem P1 of symbolic design.

sequential circuit

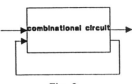

Fig. 2.

Interconnected circuits

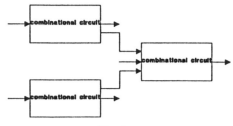

Fig. 3.

We describe in detail in Section IV how to compute an encoding of the input and output symbols that is compatible with a minimal table. Such an encoding allows us to transform the minimal symbolic cover into a Boolean representation with as many product terms as the minimal symbolic cover. Even though this mapping is not sufficient to imply the minimality of the Boolean cover, the encoded cover can be considered a good solution to the problem. It is important to remark that the length of the encoding (i.e., the number of binary variables) needed to encode each symbol may have to be larger than the minimum length required to distinguish all the symbols in each field (i.e., the ceiling of the logarithm in base 2 of the number of elements in each field). Therefore, it is interesting to compute minimal length encodings compatible with a minimal symbolic representation and to tradeoff possibly the minimality of the number of rows in a table for the number of bits required to encode each field.

Let us now consider the design problem P4 of Section I. Any sequential circuit can be implemented by feeding back the (some of the) outputs of a combinational circuit to its inputs, possibly through a register (Fig. 2). A general model of a sequential circuit is the *finite-state machine* model; the machine is synchronous if the feedback path contains synchronous registers. Finite-state machines are generally represented by **state tables**. State tables consist of four fields related to the primary inputs/outputs and to the present/next state representation. While, in general, all these fields may be represented by symbols, it is customary to represent the primary inputs and outputs in terms of binary variables and the internal states in terms of mnemonic symbols. A classical problem is to find an optimal encoding of the state symbols that correspond to an optimal implementation. This problem is referred to as optimal state assignment [9], [10].

Optimal state assignment can be solved by symbolic design by minimizing the state table using symbolic minimization and by computing a state encoding compatible with the minimal table [5]. The feedback path makes this problem different from designing combinational units. In particular, the state symbols appear in both an input and an output column of the state table and must be encoded consistently: the set of state symbols must be encoded while satisfying the group and the ordering constraints simultaneously. The limitations and the encoding procedure is described in Section IV.

Eventually symbolic design can be applied to interconnected logic circuits. Consider two units, to be implemented by two-level logic macros, that communicate through a bus (Fig. 3). If the representation of the information is transmitted across the bus is irrelevant to the design, symbolic optimization can be used as follows. The transmitting unit can be represented by a table with a symbolic output field and the receiving unit by another table with a symbolic input field. The tables corresponding to both units are optimized by symbolic minimization and the set of symbols, representing the communication signals, can be encoded as in the previous cases. Needless to say, this method can be extended to the interconnection among any number of modules, implementing combinational or sequential functions.

III. SYMBOLIC MINIMIZATION

A. Definitions

Symbolic functions are switching functions whose variables can take a finite set of values. Each value is represented by a **word** (or mnemonic), i.e., by a string of characters. A symbolic variable s has a set of admissible values S. The symbol ϕ is reserved to denote that variable s does not take any value of S. For functions of n input variables and m output variables, let S_i^I, $i = 1, 2, \cdots, n$, and S_i^O, $i = 1, 2, \cdots, m$, be the set of admissible values for the corresponding variables s_i^I and s_i^O. Then the domain of the symbolic function is the Cartesian product $S^I \equiv S_1^I \times S_2^I \times \cdots \times S_n^I$ and the range is the Cartesian product $S^O \equiv S_1^O \times S_2^O \times \cdots \times S_m^O$. A generic element of the domain is denoted by s^I and one of the range by s^O.

A **completely specified symbolic function** of n input variables and m output variables is a function $f: S^I \rightarrow S^O$ that maps each element of the domain to an element of the range. An **incompletely specified symbolic function** is a function having the property that, for some inputs, some output variables can take any value in the corresponding range. The collection of these points of the domain is called the *don't care* set of that particular output variable.

Example 6: The truth table of Example 1 is a representation of a completely specified symbolic function with $n = 2$ inputs and $m = 1$ outputs. Here,

$$S_1^I = \{\text{DIR, IND, INDEX}\}; \quad S_2^I = \{\text{AND, OR, ADD, JMP}\};$$

$$S^O = \{\text{CNTA, CNTB, CNTC, CNTD}\}.$$

Boolean or **binary-valued** functions are symbolic functions whose variables can take the values $S = \{0, 1\}$.

The domain is $\{0, 1\}^n$. The range of a completely specified function is $\{0, 1\}^m$. For incompletely specified functions, let the symbol * represent the *don't care* condition. Then the range is $\{0, 1, *\}^m$. Similarly, the variables of **multiple-valued** functions can take the values $S = \{0, 1, \cdots, p - 1\}$, where p is the radix of the representation [17], [12]. Algebras have been developed for both the Boolean and the multiple-valued [15] representations. The representation of the result of Boolean operations are based on a **linear order** of S. Let $r: S \rightarrow N$ be an enumeration consistent with the linear order [16], where N is the set of natural numbers. Then

i) Product (AND): $s \wedge s' \equiv r^{-1} \min (r(s), r(s'))$

ii) Sum (OR): $s \vee s' \equiv r^{-1} \max (r(s), r(s'))$

iii) Complement (NOT): $\bar{s} \equiv r^{-1}(p - 1 - r(s))$.

Note that the order does not affect the semantics of the representation of a switching function; however, the order may strongly affect the size of the representation. For example, canonical representations, such as *sum of products* or *product of sums* depend on the linear order of S; in particular, the minimal representations of a Boolean function and of its complement as *sum of products* have different sizes, and algorithms have been developed to exploit this fact [19].

Representations of symbolic functions depend on the definitions of the operations among words. Unfortunately, no order relation is meaningful *a priori* among the elements of a symbolic description. For this reason, operations on symbolic representations are related to order relations among words, and appropriate order relations are introduced to obtain convenient representations of symbolic functions. In the following presentation, single-output functions are considered first, i.e., $m = 1$, to simplify the notations. The extension to multiple-output functions is then shown.

Symbolic functions are represented here in a particular canonical notation: *sum of products* or more exactly *sum of products of symbolic literal functions*. Let S be the set of admissible values for a variable s. A **symbolic literal** is a nonempty subset $\sigma \subseteq S$. For any variable $s \in S$, the **symbolic literal function** is defined as follows:

$$l(s, \sigma) \equiv \begin{array}{ll} \textbf{TRUE} & \text{if } s \in \sigma \\ \textbf{FALSE} & \text{else.} \end{array}$$

Example 7: Consider the set $S_1^I = \{\text{DIR}, \text{IND}, \text{INDEX}\}$ corresponding to the set of admissible values of the first input variable in Example 6. An example of a symbolic literal is DIR IND. The corresponding literal function is TRUE for either $s = \text{DIR}$ or $s = \text{IND}$.

By using a *sum of product of symbolic literal functions* representation, only the order in S^O affects the representation because the literal function maps words into the pair of values (TRUE, FALSE) independently of the order in S^I. Note that if a linear order relation is applied on the range, symbolic function representations in this canonical

form are equivalent to multiple valued logic function representations [17] in the same form. In particular, a multiple-valued representation can be obtained from a symbolic representation by interchanging each symbol s with $r(s)$, where $r(\cdot)$ is the appropriate enumeration.

Since an order in S^O is not necessarily given, the definitions of the representation of a symbolic function are compatible with a set, possibly empty, of **partial order relations** among the elements of the range. Let $R = \{(s, s'); s, s' \in S^O\}$ be a partial order on S^O. We say that s **covers** s' if either $s = s'$ or $s = \phi$ or $(s, s') \in TR$, where TR is the transitive closure of R [1]. The **symbolic sum** of two words s and s' is well defined only if a covering relation exists among the elements involved. In particular

$$s \vee s' = \begin{array}{ll} s & \text{if } s \text{ covers } s' \\ s' & \text{if } s' \text{ covers } s \end{array}$$

else the symbolic sum is ambiguous or ill-defined.

A **symbolic product-term** (or symbolic product) of literals is the $n + 1$-tuple $(\sigma_1, \cdots, \sigma_n, \tau)$, where $\sigma_i \subseteq S_i^I$, $i = 1, 2, \cdots, n$; $\tau \in S^O$. The word τ is called the **output-part** of the literal. A **symbolic product** $p(s', \tau)$ **of literal functions** $l_i(s_i^I, \sigma_i)$, $i = 1, 2, \cdots, n$, is a function

$$p(s', \tau) = \begin{array}{ll} \tau & \text{if } l_i(s_i^I, \sigma_i) = \textbf{TRUE}, \\ & \forall i = 1, 2, \cdots, n \\ \phi & \text{else.} \end{array}$$

Example 8: A symbolic product of the function in Example 1 is

DIR AND CNTB.

The symbolic product function takes the value CNTB when the two inputs take the values DIR and AND, respectively.

Two products p_1, p_2 **intersect** $(p_1 \cap p_2 \neq \phi)$ if $\exists s' \in S^I$ such that $p_1(s', \tau_1) \neq \phi$ and $p_2(s', \tau_2) \neq \phi$. Two sets of products P_1 and P_2 intersect $(P_1 \cap P_2 \neq \phi)$ if $\exists p_1 \in P_1$ and $\exists p_2 \in P_2$ such that p_1 and p_2 intersect. Two products are **output-disjoint** if either they do not intersect or they have the same output-part, i.e., $p_1(s', \tau_1)$ intersects $p_2(s', \tau_2)$ implies $\tau_1 = \tau_2$. A set of products is output-disjoint if the products are pair-wise output-disjoint.

Example 9: Consider the two symbolic products

DIR IND	ADD	CNTC
DIR	ADD JMP	CNTC

The two symbolic product intersect because, for the symbolic input $s' = \text{DIR ADD}$, the corresponding symbolic product functions specify a symbolic value. Since both symbolic product functions take value CNTC, the product terms are output-disjoint.

A symbolic function can be represented in a *sum of product* form; if $\forall s' \in S^I$ for which the function is specified, the operation of symbolic sum among products is well defined. In particular, such a representation always exists in the following two cases: i) for any linear order on S^O, ii) if the representation is a sum of pair-wise output-disjoint products. In the former case, symbolic sum is always well defined because a covering relation is defined between each pair of symbols in S^O. In the latter, only the symbolic sum of identical values is required.

Sum of product representations are conveniently represented in tabular forms, as a stack of product terms. A **symbolic implicant** is a symbolic product $p(s^I, \tau)$ such that $\forall s^I \in S^I$ for which the symbolic function is specified, $f(s^I)$ covers $p(s^I, \tau)$. A **symbolic cover** of a symbolic function is a set of implicants $P = \{p_1, p_2, \cdots, p_{|P|}\}$ whose sum is $f(s^I)$, $\forall s^I \in S^I$ for which the symbolic function is specified. Since symbolic sum depends on the order R on S^O, we denote a symbolic cover by the pair $C(P, R)$. The **cardinality** of a symbolic cover is $|P|$ and depends on R. A **minimum** symbolic cover of a symbolic function is a cover of minimum cardinality. A **minimal** (local minimum) symbolic cover of a symbolic function is a cover such that no proper subset is a cover of the same function.

Example 10: The following table is a symbolic cover of the function specified in Example 6:

INDEX	AND OR ADD JMP	CNTA
DIR	AND OR	CNTB
IND	AND	CNTB
IND	OR JMP	CNTD
DIR IND	ADD	CNTC
DIR	JMP	CNTC

Note that the product terms are pair-wise output-disjoint. Therefore, this representation is output-disjoint and is compatible with any set R of partial order relations on S^O, and in particular the empty set. The following table is another symbolic cover of the function specified in Example 6:

INDEX	AND OR ADD JMP	CNTA
DIR IND	AND OR	CNTB
DIR IND	ADD JMP	CNTC
IND	JMP OR	CNTD

Here, $R = \{(\text{CNTD}, \text{CNTB}); (\text{CNTD}, \text{CNTC})\}$. Note that the fourth product term has an intersection with the second and third one and these products are not output-disjoint. By choosing this particular partial order, the cover cardinality is reduced by two. Moreover, note that the first implicant can be removed by assuming that CNTA is the default output, as pointed out in Example 5.

B. Symbolic Minimization

Symbolic minimization is a procedure that attempts to determine a symbolic cover of a symbolic function in a minimum number of product terms. Finding a minimum symbolic cover is a difficult task. An analysis of the computational complexity of the problem has not been done yet. However, we conjecture that any method to find a minimum cover should involve the solution of a covering problem, which is a NP-complete problem [8]. Therefore, heuristic algorithms are used to determine a minimal (local minimum) solution. It is important to remark that re-

cent progress in heuristic logic minimization has led to techniques which very often yield minimum solutions in the binary [11], [2] and multiple-valued [18] case. We assume that the reader is familiar with heuristic logic minimization [11], [2]. Since most of the routines in the symbolic minimization algorithm are based on logic minimizer ESPRESSO-II, we refer the reader to [2] for details.

Symbolic minimization is achieved by an iterative improvement of the initial cover $C^0(P^0, R^0)$. The symbolic function is described as input by the set of products P^0, while R^0 is an empty set of ordering relations because no order relation is meaningful *a priori* among the elements of a symbolic description. Therefore, P^0 is always a set of output-disjoint products. The main idea of symbolic minimization is to generate the order R during the minimization process. The symbolic minimizer detects partial order relations that are necessary to define sums of product terms which would decrease the symbolic cover cardinality. As a result, the order relations are determined *a posteriori* by the minimizer. The output of the minimizer is a minimal cover $C(P, R)$.

Symbolic minimization is a very complex technique. To help the reader in understanding this method and the underlying principle, we first consider symbolic functions with one symbolic output only and we present a simplified version of the main loop of the symbolic minimization algorithm. Then the complete algorithm is described.

1) A Simplified Symbolic Minimization Loop: Symbolic minimization can be achieved by an iterative loop that uses a multiple-valued-input, binary-valued-output minimization procedure. This procedure can be regarded as a black box that takes as input the representation of a multiple-valued-input function and its *don't care* set and returns a minimal representation. Computer programs ESPRESSO-II [2], ESPRESSO-MV [18], and MINI [11] can be used in this regard.

It is convenient to represent the partial order R by a directed acyclic graph $G(V, A)$, where the vertex set V is in one-to-one correspondence with S^O. The edge set A is initialized empty and is constructed during the minimization process. An edge between two vertices defines an order relation between the corresponding elements of S^O. Therefore the sum of two distinct elements of S^O is well defined if there is a directed path between the corresponding vertices.

Let us arbitrarily label the elements of the range: $S^O = \{s_i^O, i = 1, 2, \cdots, q\}$. Let ON_i, $i = 1, 2, \cdots, q$, be the subset of the initial set of product terms P^0 consisting of the product terms whose output part $\tau = s_i^O$. Note that the set ON_i does not intersect the set ON_j if $i \neq j$ because P^0 is output-disjoint. Each set ON_i specifies a symbolic single-output single-valued function. The original symbolic function can be seen as a collection of q multiple-valued-input, binary-valued-output functions whose *on* set corresponds to the points of the domain mapped into s_i^O, whose *off* set corresponds to those points mapped into s_j^O, $j \neq i$, and whose *don't care* set corresponds to the unspecified points [2]. A representation of each set ON_i

by a minimal number of product terms, denoted here by M_i, can be obtained $\forall i = 1, 2, \cdots, q$ by using a multiple-valued-input, binary-valued-output minimization technique. In principle, by performing q minimization in this way, a minimal cover of the original function can be computed as $P = \cup_{i=1}^{q} M_i$, i.e., as a collection of the q minimal covers M_i. It is shown in [2] how to perform the q minimizations simultaneously. This procedure is called here **disjoint minimization** because the minimal cover P of any completely specified function is output-disjoint.[2] Disjoint minimization does not exploit the benefit of choosing an order R to minimize the cover, and therefore is a weak optimization technique.[3]

The main idea of symbolic minimization is that the cover P is not constrained to be output-disjoint by introducing appropriate order relations among the elements of S^O. For example, suppose that $(s_j^O, s_i^O) \in R$. Then any point of the domain represented by ON_j can be used to reduce the cardinality of ON_i in the minimization process. In the minimal symbolic representation, such point is still mapped into s_j^O because $(s_j^O, s_i^O) \in R$. In other words, the subset of the domain represented by ON_j is a part of the *don't care* set while minimizing ON_i. We represent the *don't care* set by the set of product terms DC_i. In this case, $ON_j \subseteq DC_i$.

Example 11: Consider the first cover of Example 10. Let $s_i^O =$ CNTB and $s_j^O =$ CNTD. Then, ON_i is

DIR   AND   OR   CNTB
IND   AND   CNTB

Suppose (CNTD, CNTB) $\in R$. Then DC_i includes

IND   OR   JMP   CNTD.

Therefore, the point of the domain $s^I =$ IND OR can be used to reduce the cardinality of ON_i that can be represented as

DIR IND   AND   OR   CNTB.

To minimize ON_i, an explicit representation of the corresponding *don't care* set is needed. Equivalently, the *off* set can be specified and the *don't care* set obtained by complementation of the *on* and *off* sets. To take advantage of the order relations, we use a definition of the *off* set different from that used in [2] and mentioned before. For our purposes, the *off* set corresponding to ON_i is the subset of s^I that is mapped by the function f to a value different than s_i^O and covered by s_i^O, because $\forall s^I \in s^I$ s.t. $f(s^I) \neq s_i^O$ and $f(s^I)$ is covered by s_i^O, $p(s^I, s_i^O)$ is not an implicant of the function. If $G(V, A)$ represents the partial order, then the *off* set can be defined as a set of product terms: $OFF_i = \cup_J ON_j; J = \{ j$ s.t. $\exists$ a path from v_i to v_j in $G(V, A)\}$.

---

[2]If the original function is incompletely specified, the minimal cover P is still output-disjoint if we restrict the definition of intersection among implicants by considering S as the *care* set of the function.

[3]In a previous paper [5], the author used the concept of symbolic minimization to deal with the optimal state assignment problem. Since no symbolic minimization technique was available at that time, disjoint minimization was used instead.

At each iteration of the symbolic minimization loop, M_i is obtained by minimizing ON_i, using a routine that performs multiple-valued-input binary-valued-output minimization. We invoke the minimization routine with the pair (ON_i, OFF_i) so that the corresponding *don't care* set DC_i, computed by the minimizer by complementation, includes by construction of the sets ON_j for which no path exists in $G(V, A)$ from v_i to v_j. As a result, minimization may be very efficient in reducing the cardinality of ON_i because of the particularly advantageous *don't care* set. If M_i intersects ON_j, the relation (s_j^O, s_i^O) is recorded by adding (v_j, v_i) to the edge set of the graph.

SYMBOLIC MINIMIZATION LOOP
Data ON_i, $i = 1, 2, \cdots, q$;
Data $G(V, A)$;
$A = \phi$; $P = \phi$;
for ($k = 1$ to q){
    $i =$ **select** (k);
    $OFF_i = \cup_J ON_j$; $J = \{ j | \exists$ a path from v_i to v_j
                                            in $G(V, A)\}$;
    $M_i =$ **minimize** (ON_i, OFF_i);
    $A = A \cup \{(v_j, v_i)$s.t.$M_i \cap ON_j \neq \phi\}$;
    $P = P \cup M_i$;
};

Procedure **select** sorts the sets ON_i according to a heuristic criterion. Procedure **minimize** is a call to a multiple-valued-input binary-valued-output minimizer. The algorithm generates a set of symbolic products $P = \cup_{i=1}^{q} M_i$ and the directed graph $G(V, A)$.

Theorem 1: The graph $G(V, A)$ generated by the symbolic minimization loop is acyclic.

Proof: By construction. Initially, the graph is acyclic because the edge set A is empty. Suppose that at iteration k the graph has no cycles. Then at iteration $k + 1$ the graph has no cycles according to the following argument. Let i be the index returned by **select** at iteration $k + 1$. Let $\bar{J} = \{ j$ s.t.$M_i \cap ON_j \neq \phi\}$. Since the edges (v_j, v_i); $j \in \bar{J}$ are those and only those added to the edge set at iteration $k + 1$, then any cycle must include a vertex v_j, $j \in \bar{J}$ and a directed path must exist between v_i and that vertex. Let $J = \{ j | \exists$ a path from v_i to v_j in $G(V, A)\}$. Then $\bar{J} \cap J = \phi$, because $M_i \cap OFF_i = \phi$, i.e., the minimal cover of the *on* set cannot intersect the *off* set. Then no cycle can be introduced at step $k + 1$. Therefore, the final graph $G(V, A)$ is acyclic. ∎

Since $G(V, A)$ is acyclic, R represents a partial order relation on S^O. It is now important to show that $C(P, R)$ is a cover of the symbolic function specified by the initial cover $C^0(P^0, R^0)$. We assume the correctness of procedure **minimize** in returning the minimal covers M_i of the covers ON_i, $i = 1, 2, \cdots, q$.

Theorem 2: $C(P, R)$ is a minimal cover of the original symbolic function represented by $C^0(P^0, R^0)$ and $|P| \leq |P^0|$.

Proof: Consider each symbolic output value s_i^O, $i = 1, 2, \cdots, q$. Since the elements of the symbolic function domain mapped by f into s_i^O are represented by ON_i in P^0

and M_i is a minimal representation of ON_i, then for each element s' in the domain mapped by f into s_i^O and $\forall i$, $i = 1, 2, \cdots, q$, there exists at least one symbolic implicant $p \in P$ whose output part is s_i^O and s.t. $p(s', s_i^O) = f(s')$. For a generic element s' of the domain, let $P(s') = \{p \in P$ s.t. $p(s', \tau) \neq \phi\}$. Since $C(P, R)$ is not necessarily output-disjoint, the output parts of the product terms in $P(s')$ may be conflicting. However, since $M_i \cap ON_j \neq \phi$ implies $(v_j, v_i) \in A \; \forall i, j$, then the sum of the product terms in $P(s')$ in $f(s')$. Moreover, $C(P, R)$ is minimal because the covers M_i, $i = 1, 2, \cdots, q$ are minimal and $P = \cup_{i=1}^{q} M_i$. Eventually, since $|M_i| \leq |ON_i|$, $i = 1, 2, \cdots, q$, then $|P| = \Sigma_{i=1}^{q} |M_i| \leq \Sigma_{i=1}^{q} |ON_i| = |P^0|$. ∎

The order R depends on the heuristic sorting of the sets ON_i, done by procedure **select**. As a result, the cardinality of the cover $C(P, R)$ generated by the algorithm strongly depends on this routine. Several heuristics have been tried. Note that as more edges are added to the graph, it is more likely that the *off* sets become large and the *don't care* sets are small. An effective heuristic is to sort the sets ON_i in descending order of cardinality, so that the largest sets will benefit from large *don't care* sets. A key ingredient for an effective reduction of the cover minimality is that the graph should be kept as sparse as possible by introducing only the ordering relations needed to reduce the cover cardinality. Keeping the graph sparse corresponds to keep many degrees of freedom to order S^O in the later iterations of the algorithm. Note that if **minimize** is a "standard" minimization algorithm, it aims at reducing both the product of literal cardinality in each ON_i. Therefore, the local optimization of the number of literals in the **minimize** procedure may introduce new edges in the graph and reduce the likelihood of reducing the symbolic cover cardinality at a later step. For these reasons, the symbolic minimization loop has to be modified for efficiency by tuning the **minimize** routine to the symbolic minimization problem.

2) The Symbolic Minimization Algorithm: We consider here symbolic functions with multiple outputs (i.e., $m \geq 1$): one output variable is a symbolic q-valued variable; the other output variables can take values in the ordered set of symbols $\{0, 1\}$, i.e., are binary-valued variables. In this case, the symbolic implicant (product term) output part τ has m scalar components τ_l, $l = 1, 2, \cdots, m$; τ_1 can take any of the q values in $S_1^O = \{s_{1,1}^O, s_{1,2}^O, \cdots, s_{1,q}^O\}$; τ_l, $l = 2, 3, \cdots, m$ can be either 0 or 1; in addition $\tau_1 = \phi$ if the implicant does not carry any information regarding the q-valued symbolic output. This special case is considered because it applies to finite-state machines whose next-state function is specified by symbols and whose primary outputs are binary-valued functions. The general case of m symbolic valued outputs will be mentioned later.

To obtain an efficient algorithm for symbolic minimization, it is necessary to "open the black box" and examine more carefully the operations that the procedure **minimize** performs. We call **optimize** the modified procedure for multiple-valued-input minimization used in the algorithm. The symbolic minimization algorithm is as follows:

SYMBOLIC MINIMIZATION

Data $C(P^0, R^0)$;
Data $G(V, A)$;
$A = \phi$; $P = \phi$;
for $(k = 1$ to $q)$
  $\quad ON_k^0 = $ **slice** $(P^0, s_{1,k}^O)$;
$OFF_b = $ **get__off__set** (P^0);
$P^1 = $ **disjoint__minimize** (P^0);
for $(k = 1$ to $q)$
  $\quad ON_k^1 = $ **slice** $(P^1, s_{1,k}^O)$;
$P^\phi = $ **slice** (P^1, ϕ);
for $(k = 1$ to $q)\{$
  $\quad i = $ **select** (k);
  $\quad OFF_i = OFF_b \cup \cup_J ON_j^0$; $J = \{j | \exists$ a path from v_i
  $\qquad\qquad\qquad\qquad\qquad\qquad\qquad$ to v_j in $G(V, A)\}$;
  $\quad M_i = $ **optimize** (ON_i^1, OFF_i);
  $\quad P = P \cup M_i$;
$\}$;
$P = $ **merge** (P, P^ϕ);

The first improvement over the simplified symbolic minimization loop can be explained in terms of preprocessing of the input data. As mentioned before, the ordering relations R need only be introduced when they reduce the cardinality of P. Therefore, it may be useful to perform a disjoint minimization before entering the symbolic minimization loop. Disjoint minimization is done by procedure **disjoint__minimize**, which invokes a multiple-valued-input minimizer, such as ESPRESSO-II. The minimized cover P^1 is a minimal representation of the m scalar components of the function. Now we call ON_i^1, $i = 1, 2, \cdots, q$, the sets of product terms in P^1 with $\tau_1 = s_{1,i}^O$. The procedure that returns the sets ON_i^1 from P^1 is called **slice**. The sets ON_i^1 contain all the points of the domain mapped into $s_{1,i}^O$ and possibly some points of the *don't care* set. Note that there may be a nonempty subset of products in P^1 with $\tau_1 = \phi$; these product terms do not carry any information related to the first scalar component and do not affect the symbolic minimization loop: they are stored in set P^ϕ. Similarly, while the *off* sets related to each symbolic value of the first scalar component are defined in the loop, the *off* sets related to the other components do not change. We call OFF_b the union of the *off* sets corresponding to the binary-valued components of the function, i.e., components $l = 2, 3, \cdots, m$. Procedure **get__off__set** extracts the *off* set OFF_b from the original representation P^0. In the main loop, OFF_i is the union of two subsets: OFF_b and $\cup_J ON_j^0$, which represents the *off* set of the value of the first component being considered. Note that the sets ON_j^0 are a "slice" of the original representation $C^0(P^0, R^0)$, so that they correspond to the points of the domain, and only those points, that have to be mapped by f into $s_{1,i}^O$, $i = 1, 2, \cdots, q$.

In the symbolic minimization loop, procedure **select** sorts the sets ON_i^1 in descending order of cardinality. Then the sets M_i are obtained from ON_i^1 by procedure **optimize**.

Procedure **optimize** sets $M_i = ON_i^1$ and then performs operations on M_i to reduce its cardinality. Procedure **optimize** differs from procedure **minimize** (described in the previous subsection) because the intersections between M_i and ON_j^0, $j \neq i$, are monitored during the minimization process. In particular, if at any given point $M_i \cap ON_j^0 = \phi$ and a point of the domain corresponding to ON_j^0 is useful as a *don't care* point to reduce the number of implicants of M_i, this point is used and the order relation (j, i) is recorded by adding (v_j, v_i) to the edge set of the graph. If $M_i \cap ON_j^0 \neq \phi$, then the points of ON_j^0 can be used unconditionally as *don't care* points. By doing this, we introduce a new order relation only when the cardinality of M_i is reduced by deleting one (or more) implicants. Note that by monitoring the intersections $M_i \cap ON_j^0$, we avoid the set intersections after the minimization step, as done in the simplified loop. Procedure **optimize** is implemented by a modified version of program ESPRESSO-II.[4] At the exit of the main loop, procedure **merge** appends to P the implicants of P^ϕ that are not covered by (or that cannot be merged with) implicants of P.

The primary goal of symbolic minimization is to reduce the cardinality of the cover. An interesting question is to explore the role of the symbolic literals and how the secondary goal of symbolic minimization can be related to the literals. A clue can be obtained by relating symbolic minimization to the binary encoding problems sketched in Section II. The minimal symbolic cover is an intermediate step in the process of computing a minimal Boolean cover. Therefore, the symbolic literals in a symbolic cover should be chosen to ease the encoding of the minimal symbolic cover into a Boolean cover as much as possible. Note first that a literal consisting of all admissible values for the corresponding variable is equivalent to a *don't care* condition on that variable. Such a literal is called a **full** literal. Therefore, it is always convenient to attempt to expand each literal to full. This is the equivalent to attempting to minimize input literals in standard Boolean minimization. When it is not possible to expand a literal to full, it is questionable which is the optimal cardinality for each literal, i.e., the optimal number of words in each literal. The strategy used in symbolic minimization is the following. For each implicant of the minimal cover, we compute an **expanded implicant**, whose literals have maximal cardinality. The expansion process increases the literal cardinality of a product term, while retaining it as an implicant of the function. Then we compute a **reduced** implicant whose literals have minimal cardinality. The reduction process decreases the literal cardinality of a product term, while making sure that the reduced product term and the remaining ones are still a

cover of the function [2]. By comparing each expanded implicant with the corresponding reduced implicant, we can detect the *don't care* words in each literal. Such words represent input conditions that do not affect the value of the minimal cover and that can be used effectively to ease the encoding problem, as shown in Section IV. Procedure **optimize** implements the literal optimization strategy.

We show now that $C(P, R)$ is a cover of the symbolic function specified by $C^0(P^0, R^0)$. We assume the correctness of procedures **disjoint-minimize** and **optimize** in returning the corresponding minimal covers. We note first that the graph $G(V, A)$ constructed by the algorithm is acyclic because the structure of the main loop and the steps that construct the graph are the same as in the symbolic minimization loop of the previous subsection and, therefore, Theorem 1 applies.

Theorem 3: $C(P, R)$ is a minimal cover of the original symbolic function represented by $C^0(P^0, R^0)$ and $|P| \leq |P^1| \leq |P^0|$.

Proof: Let $C_1(P, R)$ be the cover of the first component of the function generated by the algorithm. Let $C_1^0(P^0, R^0)$ be the original cover of the first component. As far as the first component is concerned, the symbolic minimization algorithm is equivalent to applying the simplified loop on the sets ON_i^1, $i = 1, 2, \cdots, q$. Since all the points of the domain mapped into $s_{1,i}^0$ are represented by ON_i^1, $i = 1, 2, \cdots, q$, by Theorem 2, $C_1(P, R)$ is a minimal cover of the first component of the function specified by $C_1(P^0, R^0)$. Let now P_l, $l = 2, 3, \cdots, m$, be the cover of component l computed by the algorithm, P_l^1, $l = 2, 3, \cdots, m$, be the cover of component l after disjoint minimization, and P_l^0, $l = 2, 3, \cdots, m$, be the original cover of component l. Let $ONE_i = \{s^l \in S^l$ s.t. $f_l(s^l) = 1\}$ and $ZERO_l = \{s^l \in S^l$ s.t. $f_l(s^l) = 0\}$, $l = 2, 3, \cdots, m$. Then, for each component $l = 2, 3, \cdots, m$, P_l^1 represents all the points in ONE_l and none of the points in $ZERO_l$ by the assumption of correctness of procedure **disjoint__minimize**. Similarly, P_l represents at least all the points represented by P_l^1 and none of the points in $ZERO_l$ because P is constructed as a union of P^ϕ and the sets M_i, and no product term in M_i, $i = 1, 2, \cdots, q$, with $\tau_l = 1$ represents any point in $ZERO_l$ because M_i is obtained by **optimize** (ON_i^1, OFF_i) and $OFF_b \subseteq OFF_i$ represents the *off* set of components $l = 2, 3, \cdots, m$. Therefore, $C(P, R)$ is a cover of the original symbolic function represented by $C^0(P^0, R^0)$. The cover $C(P, R)$ is minimal because the covers M_i, $i = 1, 2, \cdots, q$, are minimal and only the products P^ϕ that are not covered by $\cup_{i=1}^q M_i$ are appended to it at the exit of the main loop. Moreover, since $|M_i| \leq |ON_i^1|$, $i = 1, 2, \cdots, q$, and $|ON_i^1| \leq |ON_i^0|$, $i = 1, 2, \cdots, q$, then $|P| = |P^\phi| + \Sigma_{i=1}^q |M_i| \leq |P^\phi| + \Sigma_{i=1}^q |ON_i^1| = |P^1| \leq \Sigma_{i=1}^q |ON_i^0| = |P^0|$. ∎

The symbolic minimization algorithm invokes q times procedure **optimize**, whose computational complexity is similar to that of minimizer ESPRESSO-II. Therfore, the total computational complexity grows linearly with the number of elements in S_1^O. The minimization procedure is

---

[4]Procedure EXPAND of program ESPRESSO-II is modified as follows. By using the terminology of [2], let EXPAND1 be the routine that takes an implicant c and a cover and returns an expanded implicant c^+ and the set W of implicants covered by c^+. Just after the call to EXPAND1, an additional routine checks the intersections $c^+ \cap ON_i^0$, $i \neq j$. If ∃ j such that $c^+ \cap ON_i^0 \neq \phi$ and $(v_j, v_i) \notin A$ and $W = \phi$, then c^+ is replaced by the original cube c.

TABLE I
RESULTS OF SYMBOLIC MINIMIZATION

Example	Original cover cardinality	Minimal cover cardinality	Minimal disjoint-cover cardinality
EX1	24	9	16
EX2	91	49	57
EX3	170	78	78
EX4	11	5	8
EX5	25	10	11
EX6	24	17	24
EX7	115	89	94
EX8	107	57	92
EX9	184	106	115
EX10	16	14	15
EX11	166	102	111
EX12	49	11	12
EX13	25	10	11
EX14	20	8	13
EX15	56	23	24
EX16	32	15	16
EX17	108	46	55
EX18	32	17	18
EX19	14	9	10
EX20	30	22	23

a heuristic procedure: no theoretical bounds on the computational complexity have been proven for heuristic minimization; however, experimental results have shown that it is practical to minimize a wide range of logic functions [11], [2] and, therefore, the symbolic minimization algorithm can be used in this perspective.

As a final remark, symbolic minimization can be extended to multiple-output functions ($m > 1$) with m symbolic outputs by extending the definitions and operations appropriately. In this case, m partial orders on the sets S_i^O, $i = 1, 2, \cdots, m$, have to be recorded. The symbolic minimization algorithm can be extended to cope with this case. However, since each symbolic implicant may imply more than one output condition, it is very complex to determine the best sequence to apply the **optimize** procedure. For this reasons, further investigation is still needed to solve the general case.

C. Implementation and Results

The symbolic minimization algorithm has been implemented in a computer program called CAPPUCCINO because it is based on the logic minimizer ESPRESSO-II. CAPPUCCINO is written in APL and incorporates a modified version of the ESPRESSO-II original program [2]. CAPPUCCINO implements the symbolic minimization algorithm described in Section III-B-2. The simplified loop was presented in Section III-B-1 to ease the understanding of symbolic minimization. It was implemented only in an early experimental stage and then superseded by the complete algorithm. CAPPUCCINO has been tested on several examples. Table I summarizes the results.

The first two numeric columns show the original and final cover cardinality. The last column shows the final cardinality obtained by disjoint-minimization, by using program ESPRESSO-II. In some cases (EX1, EX2, EX8, EX17, $\cdots$), CAPPUCCINO does significantly better than ESPRESSO-II in reducing the cover cardinality. In

some others, the advantage of introducing covering relations among the output symbols is not a major factor in reducing the cover cardinality. These comparisons have to be understood with caution because we are comparing different minimization techniques and not program performances.

Computing time ranges from a few seconds to about 20 minutes for the largest example on an IBM 3081 computer. Note that APL is interpreted and, therefore, the execution is much slower with comparison to compiled code programs. Today, CAPPUCCINO is limited to covers of about 2000 symbolic product terms due to memory limitations of the APL workspace and computing time. A compiled code implementation of the algorithm based on the data structure and the routines of program ESPRESSO-MV [18] (in place of the APL version of ESPRESSO-II) would definitely increase the capability and the performance of the program.

IV. ENCODING PROBLEMS AND ALGORITHMS

A. Encoding Problems

Symbolic minimization is used as an intermediate step in solving problems P1–P4 of Section I. Since the final result must be a binary-valued logic circuit implementation, the symbolic representation has to be translated into a binary-valued (Boolean) one. If a multiple-valued circuit implementation technology were available (including logic gates implementing the literal function [17]), then the (minimal) symbolic representation could be mapped into a multiple-valued representation with the same cardinality by interchanging the words s with $r(s)$, where $r(\cdot)$ is an appropriate enumeration. If a partial-order relation exists on a set of words, then the enumeration must be consistent with it.

Let us consider first problems P1–P3. The goal of the following encoding technique is to find a binary-valued *sum of products* representation of the switching function with as many product terms as the (minimal) symbolic representation.[5] To construct such a Boolean cover, it is sufficient to determine: 1) an encoding of the words related to each symbolic input variable such that each symbolic implicant can be represented by one Boolean implicant; 2) an encoding of the words related to each symbolic output variable that preserves the covering relations; i.e., such that the encoding of the sum of any subset of symbolic products is the sum of the corresponding Boolean products. For this reason, we consider two encoding problems: the former is related to the encoding of the

---

[5]Unfortunately, it is not possible to state the minimality of the Boolean cover because a Boolean implicant may be covered by the sum of two or more implicants. Symbolic minimization detects pair-wise covering relations and neglects (in the version presented here) one-to-many relations that cannot be expressed by the partial order (e.g., a word may be made equivalent to the simultaneous assertion of two or more of words). An extension of symbolic minimization to cope with this situation would be of great theoretical interest, but would probably also complicate the problem of finding a Boolean encoding. In practice, the Boolean covers obtained by the present technique are often minimal or close to minimal.

symbols representing the input variables; the latter to the encoding of the output variables.

Let us consider first the encoding of the input variables. If a variable can take at most two symbolic values, it has a trivial Boolean encoding. In the general case, a variable that can take more than two symbolic values is represented by more than one binary-valued variable. Then, to achieve the goal of encoding each symbolic implicant by one Boolean implicant, we must represent each symbolic literal by one product of Boolean literals. A product of Boolean literals is called **cube** or **face** because it is a subspace of the Boolean hyperspace that can be represented by a hypercube. Therefore, the encoding of the words must be such that each symbolic literal can be represented by a face (Boolean cube) that is a subspace of the Boolean space that contains the encoding of all and only the symbols in the literal [5]. If *don't care* words are specified in that literal, then it is indifferent whether the face representing the literal contains the encoding of *don't care* words or not.

The problem of encoding the words related to the output variables is different because the output part of the symbolic implicants corresponding to an output variable consists of one word only (and not of a symbolic literal with possibly more than one word, as in the case of the input variables). However, the encoding of the words related to the output variables must be such that the covering relations are preserved while transforming the symbolic cover into a Boolean cover. Therefore, the encoding must be such that, for any two words joined by an order relation, the corresponding encoding are linked by a covering relation, i.e., the first word covers bit-wise the second word.

The encoding problem derived from problem P4 has an additional constraint. In this case, there is one set (or more sets) of words corresponding to both input and output variables. The encoding of this set of words must satisfy the requirements for the encoding of the input variables and the output variables simultaneously.

We now formally present the encoding problems. Let S be a set of words to be encoded and let $n_s = |S|$. Let n_p be the cardinality of the (minimal) symbolic cover. Let n_b, the encoding length, i.e., the number of Boolean variables used to represent S. The encoding problem is studied using matrix notation. Some matrices we consider have pseudo-Boolean entries from the set: $\{0, 1, *, \phi\}$ where * represents the *don't care* condition (i.e., either 1 or 0) and ϕ represents the empty value (i.e., neither 1 nor 0). Logical product and sum on pseudo-Boolean variables is defined as follows:

$\wedge$	0	1	*	ϕ		$\vee$	0	1	*	ϕ
0	0	ϕ	0	ϕ		0	0	*	*	0
1	ϕ	1	1	ϕ		1	*	1	*	1
*	0	1	*	ϕ		*	*	*	*	*
ϕ	ϕ	ϕ	ϕ	ϕ	·	ϕ	0	1	*	ϕ

Let S be the set of values taken by a symbolic input variable. Let us consider the set of literals in the (minimal) symbolic cover related to that variable. The **word-literal incidence matrix** A (or incidence matrix in short) is a matrix: $A \in \{0, 1, *\}^{n_p \times n_s}$

$$A = \begin{bmatrix} a_{1.} \\ a_{2.} \\ \cdots \\ a_{n_p.} \end{bmatrix} = [a_{.1}|a_{.2}| \cdots |a_{.n_s}] = \{a_{ij}\}$$

where: $a_{ij} = $

    1   if word j belongs to literal i

    *   if word j is a *don't care* word in literal i

    0   else.

Example 12: Let S be the set of operation codes in the symbolic function specified in Examples 1 and 10, i.e., $S = \{$AND, OR, ADD, JMP$\}$. Consider the minimal symbolic cover of Example 10. Then

$$A = \begin{bmatrix} 1111 \\ 1100 \\ 0011 \\ 0101 \end{bmatrix}.$$

Now let S be the set of values taken by a symbolic output variable. The **partial order adjacency matrix** $B \in \{0, 1\}^{n_s \times n_s}$ (or adjacency matrix in short) is the adjacency matrix of the graph representing the transitive closure of the partial order R. If word i covers word j, then $b_{ij} = 1$. If word i covers word j and word j covers word k, then $b_{ij} = 1$, $b_{jk} = 1$ and $b_{ik} = 1$. Since covering is a transitive relation, we represent directly all the implied covering relations by matrix B. Moreover, since it is trivial that each word covers itself, we choose not to represent it by convention, i.e., $b_{ii} = 0$, $i = 1, 2, \cdots, n_s$.

Example 13: Let S be the set of controls in the symbolic function specified in Examples 1 and 10. Consider the minimal symbolic cover of Example 10. Then

$$B = \begin{bmatrix} 0000 \\ 0000 \\ 0000 \\ 0110 \end{bmatrix}.$$

The **encoding matrix** E is a matrix $E \in \{0, 1\}^{n_s \times n_b}$

$$E = \begin{bmatrix} e_{1.} \\ e_{2.} \\ \cdots \\ e_{n_s.} \end{bmatrix} = [e_{.1}|e_{.2}| \cdots |e_{.n_b}]$$

whose rows are the encoding of the words.

Definition 1: Let $a \in \{0, 1, *, \phi\}$ and $x \in \{0, 1, *, \phi\}$. The **selection** of x according to a is

$$a \cdot x = \begin{array}{ll} x & \text{if } a = 1 \\ \phi & \text{else.} \end{array}$$

Selection can be extended to two-dimensional arrays and is similar to matrix multiplication.

Definition 2: Let $A \in \{0, 1, *, \phi\}^{p \times q}$ and $X \in \{0, 1, *, \phi\}^{q \times r}$. The **matrix pseudo-Boolean selection** is

$$A \cdot X = C = \{c_{ij}\}^{p \times r}$$

where $c_{ij} \equiv V_{k=1}^{q} a_{ik} \cdot x_{kj}$ or equivalently $c_{ij} \equiv a_{i1} \cdot x_{1j} V a_{i2} \cdot x_{2j} V \cdots V a_{iq} \cdot x_{qj}$.

Let us consider the problem of encoding the symbolic input variables first. This problem was presented in [5] for the first time. We report here the most relevant results in a more general formulation. We represent the encoding of the symbolic literals by the **face** matrix $F \in \{0, 1, *, \phi\}^{n_p \times n_b}$

encoding of word i, if word i neither belongs to the symbolic literal j nor is a *don't care* word; 2) empty values. An encoding matrix E is said to satisfy the **input constraint relation** for a given incidence matrix A if

$$\bar{F}^i \Lambda F \equiv \begin{bmatrix} \bar{f}_1^i . \Lambda f_1. \\ \bar{f}_2^i . \Lambda f_2. \\ \cdots \\ \bar{f}_{n_p}^i . \Lambda f_{n_p}. \end{bmatrix} = \Phi, \quad \forall i = 1, 2, \cdots, n_s$$

where Φ is the empty matrix, i.e., a matrix whose rows have at least one ϕ entry and therefore representing no point in the Boolean space.

Example 15: The encoding matrix of Example 14 satisfies the input constraint relation. However, if we swap the first two rows of E, the input constraint relation is no longer satisfied because the encoding of the third word ADD intersects the fourth face, or equivalently

$$\bar{F}^3 \Lambda F = (\bar{a}_{.3} \cdot e_{3.}) \Lambda (A \cdot E) = \begin{bmatrix} 0 \\ 1 \\ 0 \\ 1 \end{bmatrix} \cdot [10] \quad \Lambda \quad \begin{bmatrix} 1111 \\ 1100 \\ 0011 \\ 0101 \end{bmatrix} \cdot \begin{bmatrix} 01 \\ 00 \\ 10 \\ 11 \end{bmatrix}$$

$$= \begin{bmatrix} \phi\phi \\ 10 \\ \phi\phi \\ 10 \end{bmatrix} \Lambda \begin{bmatrix} ** \\ 0* \\ 1* \\ ** \end{bmatrix} = \begin{bmatrix} \phi\phi \\ \phi 0 \\ \phi\phi \\ 10 \end{bmatrix} \neq \Phi.$$

$$F = \begin{bmatrix} f_1. \\ f_2. \\ \cdots \\ f_{n_p}. \end{bmatrix}.$$

Each row of F is a face of the n_b-dimensional Boolean hypercube and corresponds to the face that encodes the symbolic literal. The face matrix can be obtained by performing the matrix pseudo-Boolean selection of E according to an incidence matrix A

$$F = A \cdot E.$$

Example 14: Consider the incidence matrix of Example 12 and the encoding of Example 4. Then

$$E = \begin{bmatrix} 00 \\ 01 \\ 10 \\ 11 \end{bmatrix} \quad F = A \cdot E = \begin{bmatrix} ** \\ 0* \\ 1* \\ *1 \end{bmatrix}.$$

Now let $\bar{A} = \{\bar{a}_{ij}\}$, where $\bar{a}_{ij} = 1$ if $a_{ij} = 0$; else $\bar{a}_{ij} = 0$. Then $\bar{F}^i \equiv \bar{a}_{.i} \cdot e_{i.}$ is a matrix whose rows are 1) the

The problem of encoding the values taken by a symbolic input variable is equivalent to finding an encoding matrix satisfying the input constraint relation. An optimal solution is one of minimal encoding length. Therefore, we can state

Encoding problem E1: Given an incidence matrix A, find an encoding matrix E with minimal number of columns that satisfies the input constraint relation.

Let us consider now the problem of encoding the output variables.

Definition 3: Let $A \in \{0, 1\}^{p \times q}$ and $X \in \{0, 1\}^{q \times r}$. The **matrix Boolean selection** is

$$A \circ X = C = \{c_{ij}\}^{p \times r}$$

where $c_{ij} \equiv V_{k=1}^{q} a_{ik} \Lambda x_{kj}$ and the sum (V) and product (Λ) operators on Boolean variables have the usual meaning:

V	0	1
0	0	1
1	1	1

Λ	0	1
0	0	0
1	0	1

Let now $G = B \circ E$. Row i, $i = 1, 2, \cdots, n_s$, of matrix G is the logical sum of the encoding of the words that must be covered by the encoding of word i. Therefore, we say that a matrix E satisfies the **output constraint relation** for a given adjacency matrix B if E covers G or equivalently

$$\overline{E} \wedge G = O$$

where $\overline{E}$ is the Boolean complement of E and O is the matrix of 0 entries.

In this case, the problem of encoding the values taken by a symbolic output variable is equivalent to finding an encoding matrix satisfying the output constraint relation. An optimal solution is one of minimal encoding length. Therefore, we can state

Encoding problem E2: Given an adjacency matrix B representing to a partial order, find an encoding matrix E with minimal number of columns that satisfies the output constraint relation.

The solution of problem P1 (P2 or P3) requires the solution of encoding problem E1 (E2 or both) for each symbolic variable, after symbolic minimization. The solution of problem P4 requires the encoding of one set (or more sets) or words corresponding to both input and output variables, after symbolic minimization.

Encoding problem E3: Given an incidence matrix A and an adjacency matrix B representing a partial order, find an encoding matrix E with minimal number of columns that satisfies both the input and the output constraint relation.

We explore now the existence of solutions to the encoding problems E1, E2, and E3. It was shown in [5] that 1-hot encoding satisfies always any input constraint relation.

Theorem 4: The identity encoding matrix $E = I \in \{0, 1\}^{n_s \times n_s}$ satisfies the input constraint relation for any given incidence matrix A.

Proof: It is reported in [5]. ∎

Theorem 5: Given any incidence matrix A, let $\tilde{A}$ be any Boolean matrix obtained from A by replacing any * entry by 1 or 0. Then $E = \tilde{A}^T$ satisfies the input constraint relation.

Proof: It can be derived from the proof of a similar theorem [5] in a straightforward way. ∎

Theorems 4 and 5 show that there always exist an encoding that satisfies the input constraint relation, but the length of the encoding suggested by the theorems are often far from the minimal length.

For the output variable encoding problem, there exists a trivial solution corresponding to the 1-hot encoding solution to the input encoding problem. In particular, if we enumerate the words consistently with the partial order (i.e., if we reorder the words so that word i never covers word j if $j > i$), then a strictly upper triangular matrix of 1 entries is a valid solution. Moreover, note that the first column of this encoding matrix can be dropped, because all entries are 0.

Theorem 6: The encoding matrix $E = PU$ satisfies the output constraint relation for any given adjacency matrix B representing a partial order, where $U \in \{0, 1\}^{n_s \times n_s - 1}$; $U = \{u_{ij}\}$; $u_{ij} = 1$ if and only if $j \geq i$ and P is a permutation matrix such that $P^{-1}BP$ is a strictly upper triangular matrix.

Proof: Note first that since B represents a partial order, there always exists a symmetric permutation such that $P^{-1}BP$ is a strictly upper triangular matrix. Now $G = B \circ E = B \circ PU = BP \circ U$. Then $\overline{E} \wedge G = \overline{PU} \wedge BP \circ U = P(\overline{U} \wedge W)$, where $W = P^{-1}BP \circ U$. Since $P^{-1}BP$ is a strictly upper triangular matrix, then $w_{ij} = 0$ if $j \leq i$. However, $\overline{u}_{ij} = 0$ if $j \geq i$. Then $\overline{E} \wedge G = O$, where O is a matrix of 0 entries and the output constraint relation is satisfied. ∎

Theorem 7: Given any adjacency matrix B representing a partial order, $E = \overline{B}^T$ satisfies the output constraint relation.

Proof: Let $G = B \circ \overline{B}^T$. Then $g_{ij} = V_{k=1}^{n_s} b_{ik} \wedge \overline{b}_{ik} = 0$, $\forall i, j = 1, 2, \cdots, n_s$. Therefore, $\overline{E} \wedge G = O$ and the output constraint relation is satisfied. ∎

Theorems 6 and 7 show that there always exist an encoding that satisfies the output constraint relation. As in the previous case, the length of these encodings is often far from the minimal length.

Unfortunately, it is not possible to state the unconditional existence of an encoding that satisfies both the input and the output constraint relations simultaneously.

Theorem 8: Given any incidence matrix A and any adjacency matrix B representing a partial order, a necessary and sufficient condition for the existence of an encoding that satisfies both the input and the output constraint relations is that for each triple of words $r, s, t \in S$ such that $b_{rs} = 1$ and $b_{st} = 1$, $\exists k$ s.t. $a_{kr} = 1$; $a_{ks} = 0$; $a_{kt} = 1$.

Proof: Necessity. For the sake of contradiction, suppose there exists an encoding matrix E satisfying both the input and the output constraint relation and suppose that $b_{rs} = b_{st} = 1$ and for some k, $1 \leq k \leq n_s$, $a_{kr} = 1$, $a_{ks} = 0$, and $a_{kt} = 1$. Let $J_{rt} = \{j | e_{rj} = 1 \text{ and } e_{tj} = 0\}$ and $J_{st} = \{j | e_{sj} = 1 \text{ and } e_{tj} = 0\}$. Since by assumption the encoding of r covers the encoding of s which covers the encoding of t, then $J_{rt} \subset J_{st}$. Then face f_{k} is such that $f_{kj} = *$ if $j \in J_{rt}$ and either $f_{kj} = *$ or $f_{kj} = e_{tj}$ if $j \notin J_{rt}$. Since $e_{sj} = e_{tj} \ \forall j \notin J_{st}$ and $f_{kj} = * \ \forall j \in J_{st}$ then $f_{k} \wedge e_{s} \neq \Phi$, the input constraint is not satisfied and we have a contradiction.

Sufficiency. Let $\tilde{A}$ be a matrix obtained from A by replacing the * entries by 1 or 0. Then, $\tilde{E} = \tilde{A}^T$ satisfies the input constraint relation. Let E be a matrix constructed as follows. For each column k of $\tilde{E}$, let

$$J_k^0 = \{j | \tilde{e}_{jk} = 0 \ \text{ and } \ \exists x \text{ s.t. } \tilde{e}_{xk} = 1 \ \text{ and } \ b_{xj} = 1\}$$

$$J_k^1 = \{j | \tilde{e}_{jk} = 0 \ \text{ and } \ \exists x \text{ s.t. } \tilde{e}_{xk} = 1 \ \text{ and } \ b_{jx} = 1\}.$$

The set $J_k^0(J_k^1)$ is the set of words with a 0 encoding in $\tilde{e}_{\cdot k}$ and required to be covered by (to cover) some words with a 1 encoding in $\tilde{e}_{\cdot k}$. Note that by assumption

96

$J_k^0 \cap J_k^1 = \phi$. Now let

$$e_{\cdot k} = \begin{cases} \bar{e}_{\cdot k} & \text{if } J_k^1 = \phi \\ \tilde{\bar{e}}_{\cdot k} & \text{if } J_k^1 \neq \phi \text{ and } J_k^0 = \phi \\ t_{\cdot k} & \text{else} \end{cases}$$

where $t_{\cdot k}$ is a column vector whose entries are bit-pairs

$$t_{jk} = \begin{cases} 01 & \forall j \text{ s.t. } \bar{e}_{jk} = 1 \\ 11 & \forall j \in J_k^1 \\ 00 & \text{else}. \end{cases}$$

Then $e_{\cdot k}$ satisfies the output constraint relation. Since the encoding matrix E is obtained from $\bar{E}$ by replacing some columns by their complement or by a two-bit encoding of their entries, then the encoding matrix E satisfies the input constraint relation. Moreover, since the covering relations are satisfied for each column by construction, then also the entire encoding matrix E satisfies the output constraint relation. ∎

B. Encoding Algorithms

A solution to the encoding problems E1, E2, or E3 is an encoding of minimal length that satisfies the input, output, or both constraint relations. It can be shown easily by example that, for many instances of these problems, the encoding matrices of Theorems 4–8 do not have minimal numbers of columns. Therefore, it is interesting to devise algorithms which construct a solution to problems E1, E2, and E3. Unfortunately, these are computationally complex problems of combinatorial optimization and it is not known whether an optimal solution can be computed by nonenumerative procedures. Since the growth of computation time as the size of the problem increases is a practical limitation to computer-aided design systems, we consider here heuristic algorithms for the solution of the above problems. Experimental results show that the encodings constructed by these algorithms have reasonably short length, and often equal to the minimum length solution when this is known.

The heuristic techniques presented below solves the encoding problems using greedy strategies. An encoding matrix E is grown from an initial seed matrix by concatenating rows and/or columns. At each step, the best local concatenation is computed, while keeping the previously computed encoding matrix as such. As a result, the encoding matrix grows in size, until all the rows are encodings for the words in S that satisfy the constraint relations. The heuristic selections that drive the algorithm attempt to minimize the steps that increase the number of columns to obtain a solution, and therefore guarantee a weak optimality.

There are two major approaches to constructing matrix E: a **row-based** method and a **column-based** one. In the former method, the encoding matrix is constructed row by row, i.e., by computing the encoding of the words one at a time [5]. In the latter approach, the encoding matrix is constructed column by column, i.e., by computing one bit of the encoding of all the words at each pass. This idea

was introduced first by Dolotta [7] to solve the optimal state assignment problem, though the encoding problem was stated differently.

A row-based encoding method was presented in [5] to solve problem E1. This technique is also effective for solving problem E2. The algorithm can be sketched as follows.

STEP 1: Select a word not yet encoded.
STEP 2: Determine the encodings for that word satisfying the constraint relation restricted to that word and to the previously encoded words.
STEP 3: If no encoding is found, increase the code dimension by adding a column to E and go to STEP 2.
STEP 4: Assign an encoding to the selected word by adding a row to E.
STEP 5: If all words have been encoded, stop. Else go to STEP 1.

We refer the reader to [5] for the details of the algorithm when applied to problem E1. We describe here in brief how this algorithm can be used to solve problem E2. The words are selected iteratively at STEP 1 by choosing nodes with no outgoing edges in the directed acyclic graph representing the partial-order relation. The node corresponding to the selected word is then deleted from the graph. Let n_b be the current encoding length. At the beginning, let $n_b = $ ceiling $(\log_2 n_s)$. In the first pass of the algorithm, the encoding of the first selected word is a row vector of dimension n_b with all 0 entries. In the subsequent passes at STEP 2, the encodings of length n_b that are not rows of E are determined by Boolean complementation. Encodings that do not satisfy the output constraint relation restricted to the selected and already encoded words are discarded. At STEP 3, if no valid encoding is left, a column of 0 entries is appended to matrix E. Otherwise, an encoding is selected at STEP 4 that minimizes the number of 1 entries.

The row-based encoding algorithms generate short encodings (see [5] for experimental results) but fail to be effective for large examples. In particular, the candidate encodings generated at STEP 2 may increase exponentially with the encoding length n_b. Therefore, when the encoding length increases, the computer implementation of the algorithm slows down considerably. Moreover, it is complex to handle problem E3 with this approach because the effectiveness of the algorithm depends heavily on the heuristic ordering of words at STEP 1, and for both input and output constraints there may exist conflicting orderings.

The need to handle both constraints simultaneously, as well as a desired computational-time complexity linear in n_b, has lead to the development of a column-based algorithm. In a column-based algorithm, a single-bit encoding of all the words is introduced at each step. While there always exist single-bit encodings of all the words that satisfy the output constraint relation, it is unlikely that such an encoding satisfies the input constraint relation. Therefore, we say that, given an incidence matrix A, an encoding matrix E **partially satisfies** the input constraint relation if E satisfies the input constraint relation for A', where A' is a subset of the rows of A. The **satisfaction ratio** is $n_{p'}/n_p$, where $n_{p'}$ is the row cardinality of A'. Then, the column-based encoding algorithm can be sketched as follows.

STEP 1: Select a column vector in $\{0, 1\}^{n_s}$ that satisfies the output constraint relation and corresponding to a maximal satisfaction ratio.

STEP 2: If at the first pass, let E be the selected vector. Else, append the selected column vector to E.

STEP 3: If E satisfies the input constraint relation, stop. Else go to STEP 1.

Before describing in detail the column-based encoding algorithm, we mention some properties of the encoding problems that are relevant to this method. If each column of E satisfies the output constraint relation, so does the entire matrix E and *vice versa*. This is not true for the input constraint relation. In particular, if E satisfies the input constraint relation, then a subset of columns of E may not satisfy it. However, we can prove that by appending a column to E, the satisfaction ratio cannot decrease.

Theorem 9: If E satisfies the input constraint relation for a given A, then $E' = [E|T]$ satisfies the input constraint relation, where T is any $\{0, 1\}$ matrix with n_s rows.

Proof: It is reported in [5]. ∎

Therfore, our strategy is to select columns that increase the satisfaction ratio until it reaches unity and the algorithm terminates.

Theorem 10: If E satisfies the input constraint relation for a given A, then $E' = [E|\alpha^T]$ satisfies the input constraint relation for $\begin{bmatrix} A \\ \overline{\alpha} \end{bmatrix}$.

Proof: Since E satisfies the input constraint relation for A, by Theorem 9 $[E|\alpha^T]$ satisfies the input constraint relation for A. We just need to prove that $[E|\alpha^T]$ satisfies the input constraint relation for α. Since $[\alpha^T]$ satisfies the input constraint for $[\alpha]$ by Theorem 5, then $[E|\alpha^T]$ satisfies the input constraint relation for α, again by Theorem 9. ∎

If the algorithm is used to solve problem E1, by selecting as columns the transpose of the rows of A, we construct as a solution $E = A^T$, which is a valid solution if we replace the * entries by 0's or 1's (Theorem 5). On the other hand, if the algorithm is used to solve problem E3, there exists also a column selection that increases the satisfaction ratio by at least $1/2 \times n_p$, i.e., at most two columns will be needed to satisfy the input constraint set by each row of A, while satisfying any admissible output constraint. This is shown by the proof of sufficiency of Theorem 8. However, since the optimality of the solution is measured by the number of columns of E, we need column selections that maximize the increase of the satisfaction ratio. The column selection procedure is described in the detail in the sequel.

For computational efficiency, it is important to reduce the number of rows of A to the minimal number that represents an equivalent constraint on the encoding. It is trivial that duplicate rows can be deleted, as well as rows without 0 entries or with only one 1 entry. The former set does not represent a real constraint (all encodings must be contained in the Boolean space) and the latter set requires that some encodings must contain themselves (which is a tautology). The number of rows of A can be compressed further by using the result of the following theorem.

Theorem 11: Let $\tilde{A}$ be the subset of the rows of A with no * entries. An encoding matrix E satisfies the input constraint relation for $\tilde{A}$ if and only if E satisfies the constraint relation for A', where A' is the subset of rows of $\tilde{A}$ that are not Boolean products of two or more rows of $\tilde{A}$.

Proof: The proof is reported in [5]. ∎

The number of rows of A can be reduced according to these arguments. If the number of rows after performing the reduction is *ceiling* $(\log_2 n_s)$, then $\tilde{A}^T$ is an optimal solution to E1, where A is obtained from A by replacing the * entries by 1's or 0's. In general, there may be duplicate columns in $\tilde{A}$ which would lead to duplicate encoding of the words. This is easily resolved by appending appropriate columns to E to distinguish the duplicate encodings. The problem of duplicate encodings would not exist if we consider the original incidence matrix A obtained from a minimal symbolic cover. In fact, two columns with identical entries would imply that the corresponding words are incident to the same set of literals, and therefore are indistinguishable with respect to the switching function.

The column-based encoding algorithm is based on the ideas sketched above. The input to the algorithm is the incidence matrix A, the adjacency matrix B, and the encoding options, including possibly an upper bound n_{b_max} on the numbers of columns of the encoding matrix E. Let FI and FO be two logical flags: FI is TRUE when solving problems E1 or E3 and FO is TRUE when solving problems E2 or E3.

COLUMN-BASED ENCODING ALGORITHM
```
Data A, B;
Data FI, FO, n_b_max;
n_b = 0;
if (FI) A = clean (A);
if (FI) A = compress (A);
if (FI ∧ FO) A, B = verify_constraints (A, B);
do {
    e = column_select;
    if (n_b = 0)
        E = e;
    else
        E = [E|e];
    n_b = column cardinality of E;
    if (FI) A = reduce_constraints (A);
};
while (termination criterion not satisfied);
```

Procedure **clean** records first the multiplicity of each row of A in a weight vector. Then, duplicate rows are deleted, as well as rows without 0 entries or with only one 1 entry. Procedure **compress** reorders first the rows of A as $\begin{bmatrix} \tilde{A} \\ A* \end{bmatrix}$, where $\tilde{A}$ has no * entries. Then **compress** returns $A = \begin{bmatrix} A' \\ A* \end{bmatrix}$, where A' is obtained by deleting the rows of $\tilde{A}$ that are Boolean products of two or more rows of $\tilde{A}$, as justified by Theorem 11.

Procedure **verify__constraints** is invoked if both flags FI and FO are TRUE, that is, when problem E3 is being solved. The necessary and sufficient condition for the existence of an encoding for the given constraints (Theorem 8) is checked before beginning the encoding. If conflicting constraints are found, then some constraints must be released so that it is possible to find a solution. Unfortunately, a consequence of the release of a constraint is that the encoded Boolean cover might require more implicants that the minimal symbolic cover. An optimization sub-problem is to find the set of constraints whose release would minimize the impact on the encoded cover, i.e., that would minimize the possible increase of implicants. We have found experimentally that it is effective to keep the matrix A unaltered and release the covering constraints by lowering some 1 entries of B to 0 until the modified B matrix and the matrix A satisfy the conditions for existence of an encoding.

Procedure **reduce__constraints** aims at reducing the computational burden of selecting a column by updating the matrix A at each pass of the algorithm. The rationale of the update is as follows. Let E be the encoding matrix partially constructed at a given pass of the algorithm. A partial face matrix $F = A \cdot E$ corresponds to this encoding. Then it is possible that the encodings of some words do not belong to some Boolean subspaces specified by F. The encoding of these words may be originally required not to intersect these faces and therefore this requirement is already satisfied by the partial encoding. In this case, the corresponding entries in the A matrix, that were originally 0s, can be modified to *. If a row of A does not contain any more 0 entries, then the corresponding constraint is satisfied. Equivalently, E satisfies the input constraint relation for that row of A. Since appending additional columns to E will leave the constraint satisfied, that row can be dropped from A.

Example 16: Suppose we are solving problem E1. Let

$$A = \begin{bmatrix} 0100011 \\ 1001000 \\ 0001001 \end{bmatrix}.$$

Let $E = [0100011]^T$. Then $F = A \cdot E = [10*]^T$. Then the constraint represented by the first row of A is satisfied because the first face is 1 and the encoding of the words (in positions 1, 3, 4, and 5) which must not intersect this face are all 0's. Therefore, the first row of A can be dropped from further consideration. Moreover, the second face is 0, and the words whose encoding must not

intersect this face are in positions 2, 3, 5, 6, and 7. The encoding of the word in positions 2, 6, and 7 is 1 and cannot intersect the second face. Therefore, the constraints that remain to be satisfied can be represented by matrix

$$A = \begin{bmatrix} 1*010** \\ 0001001 \end{bmatrix}.$$

Procedure **reduce__constraints** can be summarized as follows:

```
reduce__constraints
F = A · E;
for (i = 1 to n_p) {
    for (j = 1 to n_s) {
        if (a_ij = 0 and f_·i ∩ e_·j ≠ Φ)a_ij = *;
    };
};
A = clean (A);
```

The termination criterion of the column-based encoding algorithm depends on the options which are specified. In any case, the algorithm continues to append columns to E if some encodings are equal to some others. If the flag FI is set, then the algorithm terminates when E satisfies the input constraint relation. In view of the reduction of the constraint matrix done by procedure **reduce__constraints**, the algorithm terminates when matrix A has no rows left. If an upper bound is specified on the encoding length, then a sufficient condition for termination is reaching that bound. Note that, in this case, the encoding will not necessarily satisfy the constraint relations. However, care is taken in the column selection procedure so that the constructed encoding matrix E with bounded column cardinality has all rows different from each other.

We can now describe the **column__select** procedure, which is the heart of the algorithm. We consider first the cases in which no upper bound on the encoding length is specified. Let us assume that only flag FI is TRUE. The procedure aims at minimizing the encoding length, and therefore we look for a vector in $\{0, 1\}^{n_s}$, that maximizes the satisfaction ratio at each pass of the algorithm. To achieve this goal, we consider first all the pairs of rows of A. Two rows of A, $a_1.$, and $a_2.$ are said to be **compatible** if either $a_{1j} = a_{2j}$ $\forall j$ s.t. $a_{1j} \neq *$ and $a_{2j} \neq *$ or $a_{1j} = \bar{a}_{2j}$ $\forall j$ s.t. $a_{1j} \neq *$ and $a_{2j} \neq *$. It is clear that, for each pair of compatible rows, there exist a 0-1 assignment to the * entries of either one that is an encoding vector that satisfies the input constraint relation for the pair of rows.

Example 17: Let

$$A = \begin{bmatrix} *11001 \\ 00011* \\ 11110* \end{bmatrix}.$$

The first two rows of A are compatible. The encoding $[111001]^T$ satisfies the input constraint relation for these two rows.

Given a maximal set of pair-wise compatible rows,

there exists a 0-1 assignment to any of them that is an encoding maximizing the satisfaction ratio. Therefore, procedure **column__select** determines first a set of pairwise compatible rows of A, denoted by A', maximizing a figure of merit that takes into account the set cardinality and the row multiplicity (stored in the weight vector). Let J be the set of indices of the columns of A' whose entries are not all *. Clearly, only the positions in the encoding column vector specified by J are important to satisfy the constraint relation specified by A'. Procedure **column__select** takes one row in A' and assigns 0 or 1 to the * entries in the positions specified by J in such a way to maintain compatibility. (Procedure **select__i** returns a column vector given a matrix A according to this strategy.) Then, the remaining * entries can be assigned according some tie rules. In the case that only flag FI is set, the * entries are assigned first as to minimize the size of the faces (dimension of the Boolean subspaces) represented by the corresponding face matrix. In case of further ties, the assignment is done to maximize the number of 0 entries in the A matrix that will be modified to * by procedure **reduce__constraints**.

Let us now turn to the case in which both flags FI and FO are TRUE (problem E3). The column selection involves first the detection of a vector which maximizes the satisfaction ratio, as in the previous case. We call $\tilde{e}$ this vector. Then, this vector has to be transformed to satisfy the output constraint relation as well. Let $J^0 = \{ j | \tilde{e}_j = 0$ and $\exists x$ s.t. $\tilde{e}_x = 1$ and $b_{xj} = 1\}$ be the set of words with a 0 encoding in $\tilde{e}$ and required *to be covered* by some words with a 1 encoding in $\tilde{e}$. Let $J^1 = \{ j | \tilde{e}_j = 0$ and $\exists x$ s.t. $\tilde{e}_x = 1$ and $b_{jx} = 1\}$ be the set of words with a 0 encoding in $\tilde{e}$ and required to *cover* some words with a 1 encoding in $\tilde{e}$. Note that these two sets do not intersect because the matrices A, B satisfy the assumptions of Theorem 8, after having been possibly modified by procedure **verify__constraints**. There are now four possibilities:

i) $J^1 = \phi$ and $J^0 = \phi$;
ii) $J^1 = \phi$ and $J^0 \neq \phi$;
iii) $J^1 \neq \phi$ and $J^0 = \phi$;
iv) $J^1 \neq \phi$ and $J^0 \neq \phi$.

In the first two cases, there exists a 0-1 assignment of the * entries of vector $\tilde{e}$ that satisfies the output constraint relation and we set $e = \tilde{e}$. In the third case, there exists a 0-1 assignment of the * entries of the complement of vector $\tilde{e}$ that satisfies the output constraint relation and we set $e = \bar{\tilde{e}}$. In the last case, we set $e = t(\tilde{e})$, where $t(\tilde{e})$ is a column vector whose entries are bit-pairs:

$$t_j = \begin{cases} 01 & \forall j \text{ s.t. } \tilde{e}_j = 1 \\ 11 & \forall j \in J^1 \\ 00 & \forall j \in j^0 \\ e_j | e_j & \text{else} \end{cases}$$

There exists now a 0-1 assignment to the * entries of each column of e that satisfies the output constraint relation.

Moreover, any 0-1 assignment to the * entries of e satisfies the input constraint relation for the same subset of rows of A as $\tilde{e}$. The assignment of the * entries is done according to tie rules, as mentioned before.

If the algorithm is used to solve problem E2 (only flag FO is true), then the column selection is done by determining directly a vector that leads to a short encoding among those that satisfy the output constraint relation. The strategy used to select a column, having as a primary goal the determination of a short-length encoding, is similar to that used in the case an upper bound is imposed on the encoding length.

If an upper bound n_{b_max} on the encoding length is specified, then it is imperative that the rows of E are different from each other after n_{b_max} column assignments. Therefore, there may be groups of identical rows in E with at most cardinality $2^{(n_{b\_max} - n_b)}$, after having assigned n_b columns. For this reason, the column selection is done such that, if there are groups of identical rows in $[E|e]$, their cardinality is most $2^{(n_{b\_max} - n_b)}$, where $n_{b'}$ is the column cardinality of $[E|e]$. When flag FI is TRUE (or when both flags FI and FO are true), this requirement restricts the selection of the column vector e (or $\tilde{e}$). In particular, we have to consider a reduced set of rows of A in which we search for a maximal set of compatible rows. (Procedure **drop** drops the rows of A incompatible with the code length goal from further consideration in the **column__select** routine.) The assignment of the * entries is done to satisfy first the encoding-length constraint by trying to minimize the cardinality of the groups of identical rows in $[E|e]$. In case of tie, the previous rules are used.

In the case that both flags are FALSE, (or in the case that FI is TRUE, E satisfies the input constraint relation and some rows of E are identical) then **column__select** returns a column which would minimize the cardinality of the groups of identical rows in $[E|e]$. (Procedure **select__0**.) If, in addition, flag FO is TRUE, the search is limited to the column vectors satisfying the output constraint relation. (Procedure **select__o**.)

In summary, procedure **column__select** can be represented as follows:

```
column__select
if (bound on n_b) A = drop (A);
if (FI and satisfaction ratio < 1){
    e = ẽ = select__i (A);
    if (FO){
        if (J¹ ≠ φ and J⁰ = φ) e = ẽ̄;
        if (J¹ ≠ φ and J⁰ ≠ φ)e = t(ẽ);
    };
};
else {
    if (FO)
        e = selecto__o;
    else
        e = selecto__0;
};
```

100

$e = e$ with 0-1 assignment of the * entries according to the tie rules;

We summarize now the properties of the column-based encoding algorithm in a qualitative way. The algorithm constructs an encoding matrix E. If problem E2 or E3 have to be solved, each column is chosen as to satisfy the output constraint relation. Therefore, so does the entire matrix E. If problem E1 or E3 have to be solved, each column is chosen as to satisfy partially the input constraint relation, or equivalently to satisfy the input constraint relation for a subset of the rows of A. These rows are then discarded from further consideration. After a finite number of steps, matrix E satisfies the input constraint relation. The heuristic selection of the column attempts to increase maximally the satisfaction ratio at each step, and therefore guarantees a weak minimality of the column cardinality of E. If an upper bound on the encoding length is specified, then the encoding may satisfy only partially the constraint relations, but attempts to satisfy most of the constraints.

The construction of matrix E with bounded column cardinality allows to tradeoff the minimality of the cover cardinality for that of the encoding length. In particular, given an encoding of length n_b, it is possible to determine the cover cardinality (or a bound on the cover cardinality n_p) on the basis of the satisfied constraints. Therefore, for a particular switching function, it is possible to determine a set of pairs of parameters (n_b, n_p) that relate to the size of the implementation.

The worst-case computational cost of the algorithm grows cubically with n_s because the **column__select** routine involves $0(n_s^2)$ operations and is invoked $n_b = 0(n_s)$ times. It grows quadratically with n_p because of the pairwise comparisons in routines **column__select** and **reduce__constraints**. It grows linearly with n_b. Note that n_b is not an input datum parameter, but the linear growth shows that the amount of computation per column is constant.

As a final remark, we would like to compare the column-based encoding algorithm with the one proposed by Dolotta [7] and later perfected by Weiner [23], Torng [22], and Story [20]. These algorithms addressed the state assignment problem for finite-state machines, and the mechanism of the encoding algorithm is similar to the one presented here. However, the selection of the columns was based on a heuristic criterion; in particular, a scoring function was used to select a column on the basis of the likelihood that logic minimization, which would have followed the encoding, could reduce the two-level cover cardinality. (Story's method optimized the number of and–or inputs for each column choice.) In our approach, we relate each column assignment to the satisfaction of some input constraints and therefore to a known reduction of the cover cardinality. Therefore, column selection is related to cover minimality in a deterministic way. However, our encoding technique is still a heuristic one because the greedy strategy considers and assigns only one column at a time.

TABLE II
OVERALL RESULTS

Example	n_i	n_s	n_o	n_p	$n_{p'}$	n_b
EX1	2	6	2	24	11	4
EX2	7	16	7	91	49	8
EX3	1	9	2	170	78	12
EX4	2	4	1	11	6	2
EX5	2	9	1	25	10	5
EX6	1	12	1	24	17	7
EX7	7	48	19	115	89	10
EX8	8	20	6	107	68	7
EX9	11	32	9	184	107	9
EX10	1	8	1	16	14	4
EX11	9	30	10	166	103	12
EX12	4	4	4	49	11	3
EX13	2	11	1	25	10	6
EX14	4	5	1	20	8	4
EX15	8	7	5	56	23	5
EX16	8	4	5	32	15	4
EX17	4	27	3	108	49	11
EX18	4	8	3	32	17	4
EX19	2	7	2	14	9	3
EX20	4	15	3	30	22	7

C. Implementation and Results

The column-based encoding algorithm has been implemented in program CREAM. CREAM is written in APL and consists of about 25 functions. CREAM is designed to be used with CAPPUCCINO: it takes the representation of a minimal symbolic cover and generates an encoding that can replace the symbolic entries. The final result is a Boolean cover that can be implemented as a PLA (after having been folded or partitioned, if desired), or used as a starting point for multiple-level synthesis, as done by the YLE program in the Yorktown Silicon Compiler [3].

CREAM solves the encoding problem E1, E2, or E3 at the user's request. It accepts an upper bound on the number of columns to be used in the encoding. For a given bound and the corresponding encoding of all the symbolic fields, it is possible to estimate the area taken by a PLA implementation. Therefore, it is possible to estimate the area as well as the aspect ratio. This computation can be done for different bounds on the encoding, and therefore a designer can choose among several implementations with different areas and aspect ratios.

CAPPUCCINO and CREAM have been tested on several examples. Some results are reported in Table II.

The examples are sequential circuits, i.e., we are solving problem P4 by symbolic minimization first and by solving encoding problem E3 after. The symbolic representations have four fields corresponding to the primary inputs/outputs (which are represented by Booleans variables) and the present/next states (which are represented by symbolic variables). The first four numeric columns represent the parameters of the function: n_i is the number of primary inputs, n_s is the number of states, n_o is the number of primary outputs, and n_p is the cardinality of the initial symbolic cover. The last two columns show the final cardinality of the Boolean cover and the state encoding length. Note that for a few examples (As EX1, EX4, $\cdots$), the Boolean cover cardinality is larger than the minimal symbolic cardinality (reported in the third column of Table I) because it was not possible to satisfy all

TABLE III
ENCODING-LENGTH COMPARISONS

Example	n_s	$\log_2 n_s$	n_b	$n_{b'}$	$n_{b''}$
EX1	6	3	4	3	3
EX2	16	4	8	7	7
EX3	9	4	12	12	12
EX4	4	2	2	2	2
EX5	9	4	5	4	4
EX6	12	4	7	4	6
EX7	48	6	10	8	8
EX8	20	5	7	6	7
EX9	32	5	9	7	7
EX10	8	3	4	4	3
EX11	30	5	12	9	9
EX12	4	2	3	3	3
EX13	11	4	6	4	4
EX14	5	3	4	3	3
EX15	7	3	5	5	5
EX16	4	2	4	4	4
EX17	27	5	11	9	7
EX18	8	3	4	4	4
EX19	7	3	3	3	3
EX20	15	4	7	7	5

constraints in the encoding. Note also that a further reduction in cardinality may be obtained by minimizing again the Boolean covers that are not minimal. The computing time is in the order of few seconds for all these examples, on an IBM 3081 computer.

In Table III, we try to evaluate the optimality of the encoding, in terms of encoding length. The second column represents the number of words to be encoded n_s; the third represents the minimal number of bits needed to encode the words regardless of any constraint on their encoding (i.e., *ceiling* ($\log_2 n_s$)); the fourth column represents the encoding length as constructed by CREAM as a solution to problem E3. These three columns show that the encoding length computed by CREAM is close enough to the minimum length; for most of the examples *ceiling* ($\log_2 n_s$) $\leq n_b \leq 2 \times$ *ceiling* ($\log_2 n_s$). For some examples (as EX14, EX16, EX18, EX19, $\cdots$), it is possible to provide that no encoding of length inferior to n_b can be a solution of the encoding problem. The results of CREAM can be compared with those obtained by program KISS [5], which uses a row-based encoding algorithm. The length of a solution to problem E1 computed by CREAM is given in the fifth column of the table and the length of an encoding constructed by KISS in the last column. (The algorithm of KISS can solve only problem E1.) Note that, for some examples, the row-based algorithm gives a shorter encoding. However, the corresponding implementation requires more product-terms, as shown by the fourth column of Table I.

As a final remark, it would be interesting to rate the effectiveness of the symbolic design methodology by relating the experimental results to a measure of the difficulty of the examples, as in the case of channel routing. Unfortunately, it is difficult to classify the examples on an absolute scale. The examples chosen here are derived from finite state machine (FSM) tables. The present methodology appears to be effective especially for FSM's having large and sparse transition graphs.

V. CONCLUDING REMARKS AND FUTURE WORK

Combinational and sequential circuits can be optimized using the symbolic design methodology. Symbolic functions are represented by tables of symbols, which may be direct translations of hardware description language programs. Symbolic minimization allows encoding-independent optimization of switching functions, and the encoding algorithms construct a binary representation of the symbols that translate the minimal symbolic cover into a compatible Boolean cover.

The target technology of the symbolic design method is a two-level *sum of products* circuit implementation, such as a programmable logic array. However, this technique can be used in conjunciton with other implementation methodologies, such as those supported by the Yorktown Silicon Compiler [3], by mapping the optimal two-level representation into a multiple-level logic representation which fits the implementation technology.

Future work will address the extension of symbolic design to multiple-level implementations. In this case, the objective function of the optimization will consider directly multiple-level implementations without resorting to the *sum of product* model.

It is important to note that the algorithms we presented are heuristic and can still be improved. Other schemes for symbolic minimization can be tried. The symbolic minimization loop could be replaced by a multiple-valued expansion of the output parts of the symbolic implicants that takes into account the order relations. In the case of problems P4 and E3, the symbolic minimization algorithm could include directly restrictions on the operations on the symbolic cover such that the computed cover can always be encoded into a Boolean cover with the same cardinality. The encoding algorithm could be improved by combining row-based and column-based encoding techniques or by iterating or backtracking in the encoding procedure to achieve shorter encodings. Ideally, the encoding algorithm should be merged with the minimization procedure so that the entire optimization method could use the silicon area of the physical implementation as the objective function.

ACKNOWLEDGMENT

The author would like to thank R. Brayton, R. Rudell, and A. Sangiovanni-Vincentelli for many helpful discussions and D. Ostapko and J. White for reading the manuscript.

REFERENCES

[1] A. V. Aho, J. E. Hopcroft, and J. D. Ullman, *The Design and Analysis of Computer Algorithms.* New York: Addison Wesley, 1974.
[2] R. Brayton, G. D. Hachtel, C. McMullen, and A. L. Sangiovanni-Vincentelli, *Logic Minimization Algorithms for VLSI Synthesis.* Kluwer Academic Publishers, 1984.
[3] R. Brayton, N. Brenner, C. Chen, G. De Micheli, C. McMullen, and R. Otten "The Yorktown Silicon Compiler," in *Proc. Int. Symp. on Circuit and Systems* (Kioto, Japan), June 1985, pp. 391–394.
[4] G. De Micheli, R. Brayton, and A. L. Sangiovanni-Vincentelli, "KISS: A program for optimal state assignment of finite state machines," in *Proc. ICCAD* (Santa Clara), Nov. 1984.

[5] G. De Micheli, R. Brayton, and A. L. Sangiovanni-Vincentelli, "Optimal state assignment for finite state machines," *IEEE Trans. Computer-Aided Design*, vol. CAD-4, no. 3, pp. 269-284, July 1985.

[6] G. De Micheli, "Symbolic minimization of logic functions," in *Int. Conf. on Comp. Aid. Des.*, (Santa Clara), Nov. 1985, pp. 293-295.

[7] T. A. Dolotta and E. J. McCluskey, "The coding of internal states of sequential machines," *IEEE Trans. Electron Comput.*, vol. EC-13, pp. 549-562, Oct. 1964.

[8] M. R. Garey and D. S. Johnson, *Computers and Intractability*. San Francisco: W. H. Freeman, 1978.

[9] J. Hartmanis and R. E. Stearns, *Algebraic Structure Theory of Sequential Machines*. Englewood Cliffs, NJ: Prentice Hall, 1966.

[10] F. Hill and G. Peterson, *Introduction to Swtiching Theory and Logical Design*. New York: Wiley, 1981.

[11] S. J. Hong, R. G. Cain and D. L. Ostapko, "MINI: A heuristic approach for logic minimization," *IBM J. Res. Develop.*, vol. 18, pp. 443-458, Sept. 1974.

[12] S. K. Hurst, "Multiple-valued logic—Its status and its future," *IEEE Trans. Comput.*, vol. C-33, pp. 1160-1179, Dec. 1984.

[13] A. Nichols and A. Bernstein, "State assignment in combinational networks," *IEEE Trans. Electron Comput.*, vol. EC-14, pp. 343-349, June 1965.

[14] E. J. McCluskey, "Minimization of Boolean functions," *Bell. Lab. Tech. J.*, vol. 35, p. 1417-1444, Apr. 1956.

[15] E. L. Post, "Introduction to a general theory of elementary propositions," *Amer. J. Math.*, vol. 43, pp. 163-185, 1921.

[16] F. Preparata and R. Yeh, *Introduction to Discrete Structures*. Addison Weseley, 1973.

[17] D. Rine, *Computer Science and Multiple-Valued Logic*. New York: North Holland, 1977.

[18] R. Rudell and A. Sangiovanni-Vincentelli, "ESPRESSO-MV: Algorithms for Multivalued Logic Minimization," in *Proc. Custom Int. Circ. Conf.*, (Portland, OR), May 1985.

[19] T. Sasao, "Input variable assignment and output phase assignment of PLA's," *IBM Res. Rep.*, no. 1003, June 1983.

[20] J. R. Story, H. J. Harrison, and E. A. Reinhard, "Optimum state assignment for synchronous sequential circuits," *IEEE Trans. Comput.*, vol. C-21, pp. 1365-1373, Dec. 1972.

[21] S. Y. H. Su and P. T. Cheung, "Computer minimization of multivalued switching functions," *IEEE Trans. Comput.*, vol. C-21, pp. 995-1003, 1972.

[22] H. C. Torng, "An algorithm for finding secondary assignments of synchronous sequential circuits," *IEEE Trans. Comput.*, vol. C-17, pp. 416-469, May 1968.

[23] P. Weiner and E. J. Smith, "Optimization of reduced dependencies for synchronous sequential machines," *IEEE Trans. Electron Comput.*, vol. EC-16, pp. 835-847, Dec. 1967.

Automated Synthesis of Data Paths in Digital Systems

CHIA-JENG TSENG, MEMBER, IEEE, AND DANIEL P. SIEWIOREK, FELLOW, IEEE

Abstract—This paper presents a unifying procedure, called *Facet*, for the automated synthesis of data paths at the register-transfer level. The procedure minimizes the number of storage elements, data operators, and interconnection units. A design generator named *Emerald*, based on *Facet*, was developed and implemented to facilitate extensive experiments with the methodology. The input to the design generator is a behavioral description which is viewed as a code sequence. *Emerald* provides mechanisms for interactively manipulating the code sequence. Different forms of the code sequence are mapped into data paths of different cost and speed. Data paths for the behavioral descriptions of the AM2910, the AM2901, and the IBM System/370 were produced and analyzed. Designs for the AM2910 and the AM2901 are compared with commercial designs. Overall, the total number of gates required for *Emerald's* designs is about 15 percent more than the commercial designs. The design space spanned by the behavioral specification of the AM2901 is extensively explored.

I. INTRODUCTION

THE research presented in this paper is a portion of the Carnegie-Mellon University Design Automation (CMU-DA) system [8]. Using the ISPS description [5] as input, the CMU-DA system proceeds through global optimization, design style selection, data-memory allocation, physical module binding, control allocation, chip partitioning, and mask generation phases. This paper describes the result of some research in the data-memory allocation phase.

The data paths of a digital system generally contain three types of primitive elements: storage elements, data operators, and interconnection units. Given a behavioral description, the problem of data-memory allocation includes five subproblems. They are the specification of data flow and control flow, the allocation of storage elements, the allocation of data operators, the allocation of interconnection units, and the exploration of the design space.

Prior research in data-memory allocation is outlined in Section II. The ISPS description specifies the data flow and control flow. The input to the data-memory allocator is an intermediate form, named the Value Trace (VT) [13], [20] of the ISPS description. *Facet* views the VT as a code sequence. Section III contains a code sequence which is used as a running example to illustrate the synthesis procedure. Given a list of operation sequences (in some sense, this means the performance is specified), the prob-

lem of design improvement is concerned with the minimization of the number of storage elements, data operators, and interconnection units. *Facet* transforms these three minimization problems into the clique-partitioning problem. The notion of clique-partitioning is introduced in Section IV. Sections V–VII describe the formulations and present algorithms for generating solutions for each of these three problems. Sections VIII and IX discuss the implementation of an automated data path synthesizer based on the algorithms. Evaluation of designs is discussed in Section X and the evaluation criteria are used to compare automatically synthesized designs to commercial designs in Section XI. Section XII contains conclusions and suggestions for future work. Finally, details of the clique-partitioning algorithms are included in the Appendix.

II. RELATED WORK

Over the last two decades, significant effort has been devoted to the development of methodologies for the synthesis of data paths in digital systems. This section contains a selective survey of the related research which can be found in the literature.

The EXPL system [4] uses a behavioral description as the design specification and a module set for implementation of the design. The behavior is described in ISP which is a high level hardware description language. The module set used is the Digital Equipment Corporation PDP-16 Register Transfer Modules (RTM). The behavioral description of the system to be designed is compiled into an internal representation (a graph model). Based on the graph model, a number of serial-to-parallel and parallel-to-serial transformations are developed for exploring the design space. Heuristic techniques are used to limit the search process. Examples (an RTM multiplier and a controller for a conveyor–bin system) are given to illustrate the design procedure. Applying these transformations to the initial designs of these examples, improvements up to a factor of two in cost and speed are observed.

Rege [17] addresses the problem of designing a small set of high level (higher than the register transfer level and lower than the computer level) Data modules (D-modules) useful for constructing digital systems. He divides the design activity into a number of well-defined subactivities. The activities are defined by means of three spaces: the specification space, the solution space, and the space of primitive components. He first proposes a basic model for the specification space. Then he develops a set of design and analysis techniques to survey the space

Manuscript received July 18, 1984. This work was done at the Carnegie-Mellon University, Pittsburgh, PA, and was supported by the National Science Foundation under Grant ECS-8207709.

C-J. Tseng is with AT&T Bell Laboratories, Murray Hill, NJ 07974.

D. P. Siewiorek is with the Department of Electrical and Computer Engineering and the Department of Computer Science, Carnegie Mellon University, Pittsburgh, PA 15213.

IEEE Log Number 8608705.

Reprinted from *IEEE Trans. CAD of Int. Circ. Syst.*, vol. CAD-5, no. 3, pp. 379–395, July 1986.

of possible data part implementations for a given computational task. From this a number of parametrizable D-components suitable for designing flexible high level D-modules are developed. Finally, the analysis techniques and the parametrizable D-components are used to build high level D-modules. The use of these high level D-modules to build different systems to a given user specification is also demonstrated.

A distributed style data-memory allocator is described in [9]. The tasks of allocating the data-memory of distributed systems are identified. A small number of improvement techniques are discussed. However, the possible variations of an implementation for a behavioral specification are not considered.

Based on behavioral specifications of digital systems, Hafer [10] develops a mathematical model for the data part to describe the conditions and relationships which must be satisfied. The model is then extended to facilitate automated synthesis of the data part. The use of the mixed-integer linear programming technique results in optimal solutions in terms of the given constraints. The major limitation is that, even for very small specifications, the run-time of generating a design explodes rapidly with complexity.

Two data-memory allocators are described in [21]. One, named DAA, is a rule-based data-memory allocator. DAA investigates the feasibility of applying techniques in artificial intelligence to the problem of synthesizing digital hardware. The other one, named EMUCS, adopts an algorithmic approach. Given a VT description, EMUCS maps abstract data flow elements onto hardware elements in a step by step fashion. In each step tables of costs which reflect the feasibility of binding each abstract element onto each hardware element are generated. The allocator then searches these tables to find the binding that would potentially be the least costly. The algorithm tends to perform local optimization. The lack of ability to find a globally optimal solution is the major limitation of this algorithm.

A data path generator which consists of a library of procedurally defined cells, advanced graphic tools, and a composition program is described in [19]. The input to the system is a description of what registers and operators are desired in the data path. The output is the mask-level geometric layout for an actual circuit. The graphics editor is used to design primitive cells. The composition program places and interconnects the cells into a bit-sliced array. The design tool, which is tailored to a limited domain, is intended for a highly specialized form of microprocessor, one which directly embeds an algorithm (such as the Lisp interpreter of Scheme-81) on the chip. A great deal of knowledge about the particular features of the domain is embedded in the system.

A system which is concerned with the design of central processor data paths is described in [12]. Among other things, the system contains an algorithm library and a hardware library. With the system architecture, performance requirements, and maximum cost limit specified, the system generates a set of data paths and the microprogram to control the data paths as output. By searching the algorithm library and hardware library, a solution is generated by an enumeration process. Heuristics are used to limit search space.

An interactive computer aided design system developed at the IBM Thomas J. Watson Research Center is described in [7]. The system first produces a naive implementation automatically from a functional specification. It then interacts with the designer, allowing evaluation with respect to some factors, such as timing constraints, fan-in and fan-out requirements, etc. The designer can improve the design incrementally by applying local transformations until the design is acceptable for manufacturing. To find a set of transformations and an application sequence to a wide range of initial behavioral descriptions is also a part of the research goal.

The MIMOLA design system [11], [27] is intended to be an interactive design aid. The input to the system is the behavioral specification in MIMOLA, a high level procedural hardware descriptive language. The design process is divided into several steps including syntax analysis, compilation, and allocation. A behavioral specification is first transformed for maximal parallelism, and an implementation is generated to support the sequence of operations. Then, the design process is iterated to reduce the cost of the implementation by restricting the hardware available to the system. A statistical analyzer is used to provide information such as the utilization measure and the expected frequency of use of each hardware module to aid the designer in deciding if (and how) the data part must be constrained in order to meet cost objectives. Estimates of operation delays and microprogram size are given to evaluate the resultant design.

In summary, each of the above approaches has at least one of the following limitations.

1) Local optimization rather than global optimization is considered.
2) The requirements in run time and memory space suffer from combinatorial explosion.
3) The methodology is either ad hoc or tailored to some special architecture.

This paper presents a formal procedure for the synthesis of data paths. The procedure unifies the tasks of data-memory allocation. Both global optimality and execution efficiency (in time and space) are considered in the development of the methodology. Extensive experiments have verified that it produces high quality designs.

III. A Code Sequence

As indicated in Section 1, the input to the data-memory allocator is a VT description. A VT specifies all the data flow and control flow information in the original ISPS description.[1] *Facet* views the VT as a list of operations,

---

[1] The ISPS description contains *variables* from which *values* are derived. The value trace is an ordered sequence of *value* usage. The data-memory allocator assigns several values to an individual *storage element* such that the computation is still correctly carried out. The mapping between ISPS variables and storage elements is not one-to-one or onto. Indeed, storage elements may contain several ISPS variables or an ISPS variable may appear in several storage elements.

called a code sequence. A basic block is a linear sequence of operation codes having one entry point (the first operation executed) and one exit point (the last operation executed) [1]. A code sequence generally contains a number of basic blocks linked by conditional or concurrent control constructs. Table I is a code sequence which consists of a single basic block. Code statements appearing on the same line are executed in parallel by a single control step. This code sequence will be used as a running example to illustrate the synthesis procedure.

A code sequence can appear in many different forms. Different code sequences are mapped into data paths of different cost and speed. The design space can be explored by manipulating the code sequence.

The next section will introduce the notion of clique-partitioning which unifies the tasks of data-memory allocation.

IV. Clique Partitioning

The resources for data paths are storage elements, data operators, and interconnection units. Given a code sequence, the synthesis of data paths involves three tasks: assigning the variables to a suitable number of storage elements, data operators, and interconnection units. Two variables can be assigned to the same physical resource if and only if there is no conflict in the use of the two resources. Let N be an integer which is greater than two. Then, N variables can be assigned to the same physical resource if and only if each pair of these N variables does not have usage conflict. Let each variable be represented as a node in an undirected graph. The relationship of sharability can be represented by the connectivity of the two nodes. Mapping of storage elements, data operators, and interconnection units can all be formulated into the clique-partitioning problem whose mathematical definition is presented in the next paragraph.

Let G be a graph consisting of a finite number of nodes and a set of undirected edges connecting pairs of nodes. A nonempty collection C of nodes of G forms a complete graph if each node in C is connected to every other node of C. A complete graph C is said to be a **clique** [18] with respect to G if C is not contained in any other complete graph contained in G. The **clique-partitioning problem** is to partition the nodes in G into a number of disjoint clusters such that each node appears in one and only one cluster. Furthermore, each of these clusters itself forms a complete graph (clique).

Many applications require the partitioning of a graph into the minimum number of disjoint cliques. Minimization is consistent with finding the cliques in the graph one by one. However, the search for cliques in a graph has been proved to be *NP-complete*. Related research can be found in [3], [6], [14], [15], [26]. A procedure which partitions a graph into a near minimum number of cliques is given in the appendix (Algorithm 1). The procedure uses the neighborhood property (as described in the appendix) among nodes to partition a graph into a set of disjoint cliques. The time complexity of the procedure is a polynomial function of the numbers of nodes and edges in the graph [24], [25]. The procedure has been applied to a significant number of graphs and found to generate optimal partitionings. However, the neighborhood property is not a sufficient condition for finding the clique in a graph and sometimes a suboptimal solution is possible.

The goal of data-path synthesis is to minimize the number of storage elements, data operators, and interconnection units. Therefore, the problems of data-path synthesis can be formulated into the clique-partitioning problem. However, direct application of the basic clique-partitioning procedure to the data-memory allocation problem does not generate good solutions. Two other notions are necessary. One is the notion of divide-and-conquer and the other is the notion of transitive property. The interpretation of these two notions on data-memory allocation will be detailed in Sections V–VII. The mathematical interpretation of these two notions is given in the following paragraphs.

Given a graph G, each edge of G represents some kind of relationship between the two nodes. The profit of grouping some set of nodes may override the profit of grouping other sets of nodes. Assume that the edges of the graph can be classified into several categories according to the profit measure of grouping each pair of connected nodes. Then a subgraph can be constructed from those edges which belong to the same category. The generalized clique-partitioning algorithm uses these subgraphs to direct the task of clique-partitioning and avoid grouping pairs of nodes randomly. This corresponds to the notion of divide-and-conquer. The notion of transitive property evolves from the clique-partitioning procedure itself. The algorithms we developed combine two nodes at a time. Each time a pair of nodes is merged into a cluster, the profit measure of some edges may need to be updated. This is defined as the transitive property. The transitive property can be classified into two types: the loose form transitive property and the generalized transitive property. Details are given in the Appendix.

The generalized clique-partitioning algorithms proceeds in the following way. The subgraph in which pairs of nodes have the best profit measure is reduced first. Then the pairs of nodes having the next level of profit measure are collected and reduced. Repeatedly applying the procedure to the other subgraphs, the process is stopped when a subgraph of a specified category or the original graph G becomes empty. Again, details of the generalized clique-partitioning algorithms are included in the Appendix (Algorithm 3).

Having introduced the notion of clique-partitioning, the following sections will describe the *Facet* formalization for the synthesis of data paths. Examples will be used to illustrate the formulations and solutions for the synthesis of data paths.

V. Allocation of Storage Elements

As indicated in [24], it is generally beneficial to assign more than one variable to the same physical location. This section discusses the issues of minimizing the number of storage elements.

5.1. Sufficient Conditions for Combining Two Variables

A code sequence generally contains many variables. Given a set of variables, the problem of storage allocation is to combine those variables which can share a storage element. What are the sufficient conditions for combining two variables? A variable is *live* between the time of its definition and last use. A variable is *dead* between the time of its last use and the next definition. If the live periods of two variables are not overlapped, they have disjoint lifetimes. Obviously, two variables can be combined if they have disjoint lifetimes. In reality this constraint can be relaxed. Two variables A and B can be combined if their lifetimes are overlapped in such a way that one of them is used as a source and the other is used as the destination or vice versa in the same statement. In addition, the variable which is used as the source is *dead* in the next time interval, i.e., the use is a "last use." Pure data transfers are special cases.

5.2. A Procedure for Compacting Variables

If there are n variables and each pair of variables are proved to be combinable (there are $n(n - 1)/2$ different pairs), then these n variables can be assigned to the same physical location. Let the nodes of a graph be the variables and each pair of nodes which can be combined be joined by an edge. Then a graph which contains the lifetime relationships among all the variables can be constructed. Since the goal is to assign these variables to the minimum number of physical locations, this is translated into the clique-partitioning problem.

The combination of each pair of variables which are related by pure data transfers would cause these operations to be eliminated. This improvement reduces the number of control functions. If a horizontal list in the code sequence is occupied by pure data transfers, the control step can be deleted resulting in a faster implementation. To take this property into account, the reduction of the original graph is separated into two phases. In the first phase the edges which are associated with pure data transfers in some time intervals are collected to form a subgraph. Let the original graph and the subgraph be represented by G and $G1$, respectively. The edges in G and $G1$ satisfy the loose form transitive property.[2] Algorithm 3 in the appendix can be applied. Having completed the partitioning of the subgraph, Algorithm 1 is then applied to the remaining edges of the original graph.

Once the variables have been compacted, the code sequence is updated. The names which are grouped together are assigned to the same name. Operations of moving the content of a variable to itself are deleted.

[2]Let A, B, and C be three variables whose live periods do not overlap. Furthermore, pure data transfers appear between A and B as well as A and C. But, there is no pure data transfer between B and C. If the variables A and B are assigned to the same storage element, pure data transfers will appear between C and the composite variable (A, B). The same effect can be observed if the variables A and C are assigned to the same storage element. This is referred to the transitive property for storage allocation.

TABLE I
A CODE SEQUENCE

V3 = V1 + V2 ;	V12 = V1	
V5 = V3 - V4 ;	V7 = V3 * V6 ;	V13 = V3
V8 = V3 + V5 ;	V9 = V1 + V7 ;	V11 = V10 / V5
V14 = V11 and V8 ;	V15 = V12 or V9	
V1 = V14 ;	V2 = V15	

TABLE II
RESULTS OF LIFETIME ANALYSIS

L and D represent live and dead respectively.

Time	V1	V2	V3	V4	V5	V6	V7	V8	V9	V10	V11	V12	V13	V14	V15
Entry	L	L	D	L	D	L	D	D	D	L	D	D	D	D	D
1	L	L	L	L	D	L	D	D	D	L	D	L	D	D	D
2	L	D	L	L	L	L	L	L	D	D	L	D	L	D	D	D
3	L	D	L	L	L	L	L	L	L	L	L	L	L	D	D	D
4	D	D	D	L	D	L	D	L	L	L	L	L	L	D	L	L
5	L	L	D	L	D	L	D	D	D	L	D	D	D	L	L
Exit	L	L	D	L	D	L	D	D	D	L	D	D	D	D	D

5.3. Lifetime Analysis

According to the previous discussion, the lifetime analysis is an essential process for constructing the graph which identifies the constraints of storage sharing. The problem of lifetime analysis is well understood in the area of compiler design. Details are referred to [1].

5.4. Construction of the Lifetime Compatible Graph

Let the live/dead status of all the variables be represented by a lifetime list. The compatible graph is the graph consisting of all the edges which join combinable pairs of nodes. To construct a compatible graph, a complete graph which consists of all the nodes is first created. The lifetime list and the code sequence are then traced and inspected. Unless the conditions given in Section 5.1 are satisfied, the edges which join those variables which are live in the same time interval are deleted from the graph. If an edge has already been deleted, this step is ignored. Those edges which associate with pure data transfers in some time intervals are marked.

5.5. Grouping Registers into Scratch Pad Memories

Having assigned all the variables to suitable physical locations, the next step is to investigate the possibility of grouping several registers into sets of scratch pad memories. Those variables which have disjoint access time can be grouped together. This problem can also be formulated into the clique-partitioning problem.

5.6. An Example

Let the code sequence in Table I be given. Assume that the program is itself a loop. Having executed the statements in the last line, the control flow is passed back to the statements in the first line. Applying the lifetime analysis algorithm to the example, the status of these variables in each time interval is indicated in Table II.

Having derived the live/dead history of each variable, the compatible graph can be constructed. As mentioned

TABLE III
THE COMPATIBLE VARIABLE-PAIRS

(1,9)	(1,13)	(1,14)*	(2,3)	(2,5)	(2,7)
(2,8)	(2,9)	(2,11)	(2,13)	(2,15)*	(3,8)
(3,13)*	(3,14)	(3,15)	(4,13)	(5,8)	(5,11)
(5,13)	(5,14)	(5,15)	(6,13)	(7,9)	(7,13)
(7,14)	(7,15)	(8,13)	(8,14)	(9,13)	(9,15)
(10,13)	(11,13)	(11,14)	(12,13)	(12,15)	(13,14)
(13,15)					

before, a complete graph is first constructed. Using Table II, the conflict graph can be constructed in the following way. Considering the time interval defined as "1", the variables V1, V2, V3, V4, V6, V10, and V12 are live. Each edge formed by these live variables must be deleted from the graph. The nodes V2 and V3 are used as a source and destination in the same statement. In addition, the source variable is dead in the next time interval. Therefore, the edge (2, 3) is not deleted. Repeatedly applying the procedure to the entire lifetime, the resulting compatible graph is given in Table III.

Among these combinable variable-pairs, those edges which are accompanied by "*" are associated with pure data transfers in some time intervals. They are used to construct the second graph. Algorithm 3 is used to reduce these two graphs. It results in the following composite nodes: {1, 14}, {2, 15}, and {3, 13}.

Applying Algorithm 1 to the reduced graph, the variables are finally partitioned into eight clusters. They are {1, 14}, {2, 7, 9, 15}, {3, 8, 13}, {4}, {5, 11}, {6}, {10} and {12}. The variables in each of these clusters can be assigned to the same physical location. The code sequence in Table I can be refined into the form in Table IV.

There are fifteen variables in the original code sequence. They have been compacted into eight. Furthermore, the last step has been eliminated. The program can now be realized in four steps.

VI. ALLOCATION OF DATA OPERATORS

The allocation of data operators consists of two tasks. One is the combination of the same kind of operators. The other is the grouping of various kinds of operators into arithmetic and logic units. The goal is to assign these data operations to the minimum number of clusters. The problem is again formulated into the clique-partitioning problem.

6.1. The Formulation

A data operator is called an isolated operator if it is exclusively assigned to a triple statement (a statement which consists of one operation, one or two sources, and one destination). What is the effect of grouping two isolated data operators into one composite unit? If these two operations are the same, the number of operators is reduced by one. In addition, depending on the corresponding source operands and the destination variables are the same or not, the number of multiplexers and wired-broadcast trees may be increased or decreased. What is the ef-

fect of merging an isolated operator into an Arithmetic Logic Unit (ALU) or combining two ALU's? First, the buses and the gating elements connected to the input and output ports of the ALU can be shared. If the operations have common sources or destination, the original gating elements for the input or output ports of these modules can also be shared. Therefore, it is generally beneficial to merge an isolated data operator into an ALU or to combine two ALU's [24]. An important issue for the allocation of data operators is choosing an appropriate set of operators to group together.

Let two isolated operations be given. Inspecting the relationship between these two operations, there are sixteen cases [24]. These sixteen cases can be classified into eight categories. They are listed below.

G8: The operations and the three pairs of variables are all the same.

G7: The operations are different but the three pairs of variables are the same.

G6: The operations and two pairs of variables are the same. The third pair of variables is different.

G5: Two pairs of the variables are the same. The operations and one pair of variables are different.

G4: The operations and one pair of variables are the same. The other two pairs of variables are different.

G3: One pair of the variables is the same. The operations and the other two pairs of variables are different.

G2: The operations are the same. All three pairs of variables are different.

G1: The operations and all three pairs of variables are different.

All of these subgraphs satisfy the generalized transitive property.[3] For simplicity, the loose form transitive property is assumed for them. The algorithm for allocating data operators can be described as follows:

1) Create a complete graph whose nodes are indices of all the data operators. Trace through the code sequence and delete those edges connecting nodes which are used simultaneously. Identify the category of each edge.

2) Collect edges of Category 8. Use Algorithm 3 to reduce G and G8.

3) Having reduced the subgraph of Category 8, the graph G together with the subgraphs of Categories 7, 6, 5, 4, 3, 2, and 1 are reduced one by one.

6.2. An Example

Let each operation in Table IV be assigned to a specific name. An assignment is given in Table V. Using the code sequence and operator assignment, the compatible graph

---

[3] Let three data operators which do not have usage conflicts be given. The edge formed by each pair of these three operators belongs to one of the eight subgraphs. If two of these three nodes are merged, the edge formed by the remaining node and the composite node may belong to a new subgraph. This is referred to as the transitive property for the allocation of data operators.

TABLE IV
IMPROVED CODE SEQUENCE

V3 = V1 + V2 ;	V12 = V1	
V5 = V3 - V4 ;	V2 = V3 * V6	
V3 = V3 + V5 ;	V2 = V1 + V2 ;	V5 = V10 / V5
V1 = V5 and V3 ;	V2 = V12 or V2	

TABLE V
ASSIGNING OPERATOR IDENTIFIERS

V3 = V1 $+_1$ V2 ;	V12 = V1	
V5 = V3 $-_1$ V4 ;	V2 = V3 $*_1$ V6	
V3 = V3 $+_2$ V5 ;	V2 = V1 $+_3$ V2 ;	V5 = V10 $/_1$ V5
V1 = V5 and_1 V3 ;	V2 = V12 or_1 V2	

Operator Identifiers:
$$+_1 \quad -_1 \quad *_1 \quad +_2 \quad +_3 \quad /_1 \quad and_1 \quad or_1$$

Indices: 1   2   3   4   5   6   7   8

TABLE VI
G: THE EDGES OF THE ORIGINAL COMPATIBLE GRAPH

$(1,2)^1$	$(1,3)^1$	$(1,4)^4$	$(1,5)^6$	$(1,6)^1$	$(1,7)^1$	$(1,8)^3$
	$(2,4)^3$	$(2,5)^1$	$(2,6)^1$	$(2,7)^3$	$(2,8)^1$	
		$(3,4)^3$	$(3,5)^3$	$(3,6)^1$	$(3,7)^3$	$(3,8)^3$
					$(4,7)^3$	$(4,8)^1$
					$(5,7)^1$	$(5,8)^5$
					$(6,7)^3$	$(6,8)^1$

G is depicted in Table VI. The superscript integer at the right side of each edge is the category identifier of the edge.

The edge of Category 6 in G is retrieved to form the subgraph $G6$. $G6$ only consists of one edge; it is $(1, 5)$. The original graph G and the subgraph $G6$ are first reduced. The nodes 1 and 5 are combined. In the reduced graph the categories of the edges $(1, 3)$ and $(1, 8)$ are updated to 3 and 5 respectively. The reduction procedure is continued until the list of edges becomes empty. The data operators are finally grouped into three clusters. They are $\{1, 3, 5, 8\}$, $\{2, 4, 7\}$, and $\{6\}$.

VII. ALLOCATION OF INTERCONNECTION UNITS

This section discusses the issues related to the allocation of interconnection units. The procedure involves two steps: alignment of operands and allocation of the interconnection units.

7.1. Alignment of Operands

An operation may be either commutative or noncommutative. For those commutative operations, the synthesizer has the freedom of flipping the position of the two operands. If the operands of all the operations are suitably aligned, the number of interconnection units can be decreased. Let the operands of unary and noncommutative operations be collected in two sets according to their positions with respect to the operations. The operands of the commutative operations can be suitably aligned by comparing the operands with variables in these two sets.

7.2. The Formulation

Interconnection variables which are never used simultaneously can be grouped together to form buses. The goal is to group the interconnection variables into the minimum number of clusters. The problem is again formulated into the clique-partitioning problem.

To obtain a good bus style design, it is essential to minimize both the number of buses and the total number of drivers and receivers [23]. There is no profit in combining two interconnection variables which originate from different sources and connect to different sinks. On the other hand, it is generally beneficial to group those interconnection variables which originate from the same source or connect to the same sink to share a common bus. The details of the formulation are given below.

A complete graph in which nodes consist of all the interconnection variables is first constructed. Then the code sequence is traced through. In each time interval, if two interconnection variables are used simultaneously, the edge formed by these two nodes is deleted. Those interconnection variables which are associated with the same source, even when they are used concurrently, can still share a common interconnection. Therefore, when we construct the compatible graph, these entries are not deleted.

Let the compatible graph be represented by G. When the compatible graph is constructed, if two interconnection variables originate from the same source or connect to the same sink, the edge which joins these two variables is marked. All the marked edges are collected to form the second graph (named $G1$). The loose form transitive property is applicable to G and $G1$.[4] Algorithm 3 can be applied.

7.3. Refining the Initial Allocation

An initial design might have a "join-node" in which more than one bus is connected to a single input port. It is necessary to insert a multiplexer in front of the input port. The "join-node" can easily be found by checking the data paths connected to an input port. If an input port is connected to more than one bus, then the node needs to be refined.

7.4. An Example

The example is based on the code sequence in Table V and the ALU's allocated in Section 6. Inspecting the operands of the operations associated with ALU2, it is found that the positions of the two operands of the statement "$V1 = V5$ and_1 $V3$" need to be flipped. It is changed into the form of "$V1 = V3$ and_1 $V5$". Using the indices

---

[4]Let three interconnection variables i, j, and k be given. Assume there are no usage conflicts among these three variables. Furthermore, the variables i and j as well as i and k have common source or common destination. But, the variables j and k do not originate from the same source or connect to the same sink. If the variables i and j are assigned to the same bus, the variable k and the composite variable (i, j) will have common source or common destination. The same effect can be observed if the variables i and k are assigned to the same bus. This is referred to the transitive property for bus allocation.

TABLE VII
INDICES OF INTERCONNECTION VARIABLES

Source Name	Destination Name	Indexing Integer
V1	V12	1
V1	ALU1.In1	2
V2	ALU1.In2	3
V3	ALU1.In1	4
V3	ALU2.In1	5
V4	ALU2.In2	6
V5	ALU2.In2	7
V5	ALU3.In2	8
V6	ALU1.In2	9
V10	ALU3.In1	10
V12	ALU1.In1	11
ALU1.Out	V2	12
ALU1.Out	V3	13
ALU2.Out	V1	14
ALU2.Out	V3	15
ALU2.Out	V5	16
ALU3.Out	V10	17

TABLE VIII
G: LIST OF EDGES WHICH JOIN COMBINABLE INTERCONNECTION VARIABLES

[1,2]	(1,4)	(1,5)	(1,6)	(1,7)	(1,8)
(1,9)	(1,10)	(1,11)	(1,12)	(1,14)	(1,15)
(1,16)	(1,17)	[2,4]	(2,6)	(2,9)	[2,11]
(2,14)	(2,16)	(3,4)	(3,6)	[3,9]	(3,16)
[4,5]	(4,7)	(4,8)	(4,10)	[4,11]	(4,13)
(4,14)	(4,15)	(4,17)	(5,13)	[6,7]	(6,8)
(6,10)	(6,11)	(6,13)	(6,14)	(6,15)	(6,17)
[7,8]	(7,9)	(7,13)	(7,16)	(8,9)	(8,11)
(8,13)	(8,14)	(8,16)	(9,10)	(9,11)	(9,13)
(9,14)	(9,15)	(9,17)	(10,11)	(10,13)	(10,14)
(10,16)	(11,13)	(11,15)	(11,16)	(11,17)	[12,13]
(13,14)	[13,15]	(13,16)	(13,17)	[14,15]	[14,16]
(14,17)	[15,16]	(16,17)			

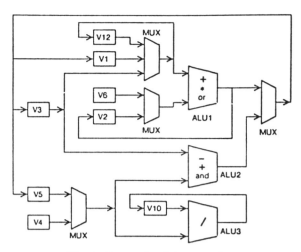

Fig. 1. Data paths of the example.

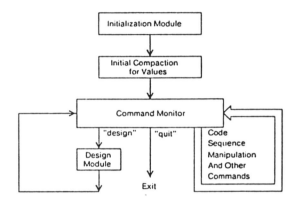

Fig. 2. Organization of the design generator.

in Table VII the compatible graph in Table VIII (*G*) is constructed. In Table VIII those edges which join interconnection variables with the same source or the same sink are enclosed by square brackets. The graph formed by these edges is called *G1*.

In *G1* nodes 14 and 16 have the maximum number of common neighbors. They are combined. *G* and *G1* are reduced. The next node to be selected should be node 15. Inspecting the nodes which are connected to node 15 and the composite node {14, 16}, it is found that both (13, 14) and (13, 15) are contained in *G*. However, among them, only (13, 15) belongs to *G1*. Since the edge (13, 15) is being deleted, the edge (13, 14) should be added into *G1*. The nodes 13 and 14 are then combined to form the composite node {13, 14, 15, 16}. This composite node forms a cluster.

Repeatedly applying the above algorithm to *G* and *G1*, the interconnection variables are finally partitioned into eight groups. They are {13, 14, 15, 16}, {1, 2, 4, 11}, {6, 7, 8}, {3, 9}, {5}, {10}, {12} and {17}. Fig. 1 depicts the completed allocation of the data-memory part.

VIII. IMPLEMENTATION

As indicated in Section I, the input to the *Emerald* design generator is the value trace. When a value trace is read in, it is represented as a hierarchical linked list. At the top level, it contains a list of VT-bodies which assume the role of procedures in other high-level languages. Each VT-body points to a list of basic blocks. Each basic block contains a two-dimensional list of operations. All the operations which are executed in the same time interval are linked in a horizontal list. A horizontal list is named a *CodeList*. The *CodeLists* are vertically chained, in order of execution, to form a basic block.

The output is an Allocated Data Path (*ADP*) which consists of five linear lists: storage elements, constants, zeros and ones for padding operations, operators, and interconnection units.

Each operation node in the code sequence contains all the information associated with the operation which includes pointers to the storage elements, data operators, and interconnection units in the *ADP*. Each operator in the *ADP* also contains a pointer back to the operation.

The code sequence representation resembles an assembly language program. The last operation in a basic block generally specifies a transfer of control. Possible transfers of control include jumping directly or indirectly to the beginning of another basic block or the beginning of a VT-body.

The structure of the *Emerald* design generator is depicted in Fig. 2. The initialization module reads in the VT file and creates internal data structures for the specification. The design generator then performs some initial compaction for values.[5] The command monitor processes commands entered by the designer. Commands are provided for manipulating the code sequence, printing the code sequence, activating the design process, etc. Once the designer specifies the code sequence, the design module can be invoked to complete a design for the code sequence. Functions of the design module include further compaction of values, packing data operations into a suitable number of ALU's or Aggregate Comparison Units (ACU's), and the allocation of interconnection units among operators and storage elements. Using the commands provided by the command monitor, the designer can generate as many designs as he/she likes. These commands can be categorized in two groups. Details will be described in the next section.

IX. Mechanisms for Design Space Exploration

The two categories of mechanisms for exploring design alternatives are manipulation of the code sequence and selection of design criteria.

9.1. Manipulation of the Code Sequence

Given a code sequence, *Emerald* creates a design for the code sequence. Alternative designs can be generated by manipulating the code sequence. As long as a modification of the code sequence does not violate data dependency among different operations, the designer can manipulate the code sequence in whatever way he/she likes.

For simplicity, the modification of a code sequence is limited to one operation at a time. There are two possible manipulations:

1) Move an operation from its current *CodeList* to another *CodeList*.
2) Serialize the code sequence by inserting a new *CodeList* between two *CodeLists* and move the operation to the new *CodeList*.

When the designer inputs a modification request, the system needs to check whether the change is allowed. Let the manipulation be limited to a basic block. Generally speaking, three conditions need to be checked:

1) Is the operation allowed to move away from its current *CodeList*?
2) Is the operation allowed to move to the new *CodeList*?
3) Is the operation allowed to move across all the *CodeLists* located between the current *CodeList* and the destination *CodeList*?

The relocation of an operation may either be in the forward or the backward direction. A forward relocation moves the operation to a *CodeList* before the current *CodeList*. A backward relocation moves the operation to a *CodeList* after the current *CodeList*. The above three conditions for moving an operation are different for forward and backward relocations.

9.2. Selection of Design Criteria

The allocation of data paths for digital systems is a complicated process; it involves making many decisions. This subsection presents some issues relevant to the selection of design criteria.

When an ISPS description is translated into a value trace, the association of a value with its ISPS counterpart is generally available. Some variables in the ISPS description are defined for input/output operations. To preserve their original roles, these variables should not be combined with other variables. *Emerald* provides the designer with the capability of specifying *reserved variables*. Those values introduced by a *reserved variable* are packed and assigned to the original variable before the compaction of other values is done. These values are not combined with other values. *Emerald* allocates a unique storage element for each reserved variable. In the resulting design, the storage element serves the same role as that defined in the behavioral specification.

Emerald treats all the array memories defined in the behavioral specification as reserved variables. The access of an array memory requires memory address registers and memory data registers to hold the address and data. Conventional von Neumann machines contain a single memory address register and a single memory data register. Processors designed recently often used multiport memory. A multiport memory has multiple address and data registers. *Emerald* provides the designers with both options. The first option is to allocate a dedicated memory address register and memory data register for an array memory. The second option is to let the system compact those values associated with memory accesses into a suitable number of memory address registers and memory data registers.

Having allocated the storage elements and constants, the process proceeds to the compaction of data operators. Assigning zero to a variable can be realized either by moving a zero to the variable or by a "clear" operation. Shifting functions can be implemented in either a barrel shifter or a shift register. Adding by one can be implemented as a counter, a local incrementer, or as part of an ALU. Similarly, subtracting by one can also be realized as a counter, a local decrementer, or as a part of an ALU. Commercial systems often use dedicated hardwares to realize special functions. Options for implementing some special functions are also provided in *Emerald*. *Emerald* asks the designer to make a decision if there is more than one choice. Experiments can be performed to investigate the effects of using different hardware.

X. Evaluation Criteria

The evaluation of designs is generally technology dependent. Even with the technology specified, it is still very difficult to estimate the exact cost and performance of a

---

[5]To relate the data flow among different basic blocks and VT bodies, the value trace contains interface values. Direct data transfers are introduced to link pairs of these values. These pairs of values, unless they are generated from formal parameters of procedure calls, can generally be combined. Therefore, they are compacted first.

design. To the best of our knowledge, evaluation models which are universally acceptable are still not available. The development of such a universal evaluation model was not a goal of this research. Instead, component counts are used to compare the cost of different designs. Several cost measures are provided. For data operators, the number of individual operators and the number of aggregate modules (the number of ALU's and ACU's) are presented. For storage elements, the number of registers and the number of bits for these registers are calculated. Since memory arrays are identical in all designs, their costs are not considered. For interconnection units, the number of multiplexers is given. The number of inputs for each multiplexing function is also specified. A multiplexer with n inputs can be realized by $n - 1$ two-input multiplexers. The total number of two-input multiplexers is calculated to compare the cost of interconnection units for each design.

Some typical TTL chips [22] are used to estimate the total number of gates required for a design. The SN74157 quadruple two-input multiplexers contains 15 gates. Therefore, each two-input multiplexer is assumed to contain 3.75 gates. The SN7476 JK flip-flop contains 32 gates for four bits. Each bit of storage element is assumed to require 8 gates. These two units will be used to calculate the number of gates required for storage elements and multiplexers. The number of gates required for aggregate data operators is separately estimated. For those aggregate data operators which contain simple operations such as AND, OR, NOT, EQL, NEQ, the number of gates is assumed to be the sum of gates required to realize these functions. For composite data operators which contain several operations, the SN74181 ALU chip is used. Delays of some TTL chips are used to evaluate the delays of some interconnection paths. Some experimental results are presented in the next section.

IX. EXPERIMENTATION

Emerald provides a fairly large number of mechanisms for exploring design alternatives. Given a VT, the number of alternative designs is almost unlimited. To validate the usefulness of the allocation procedure, the choice of an appropriate set of experiments is an important task. At this time, our major concern focuses on inspecting the quality of designs created by the allocation procedure, checking versatility of the design domain, and examining the design space. To achieve these goals, results of experiments on the ISPS descriptions for the AM2910, the AM2901, and the IBM System/370 were chosen for demonstration.[6] These three behavioral descriptions differ significantly in functions. Designs generated for these specifications are carefully investigated. Data paths produced for the AM2910 and the AM2901 are compared with the commercial designs. In cases where differences are observed, reasons for the differences are discussed. The

[6]The ISPS descriptions used in this paper were excerpted from the library of ISPS descriptions on a computer system at Carnegie-Mellon University, Pittsburgh, PA.

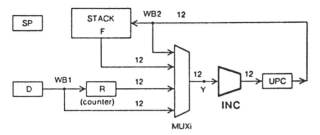

Fig. 3. Data paths of the commercial AM2910.

TABLE IX
A SEGMENT OF THE AM2910 ISPS DESCRIPTION

```
Y()<11:0> :=
    begin

    DECODE i =>
      begin
      "0 := JZ    := (Y = SP = 0; FULL = 1).
        .
      "4 := PUSH  := (Y = uPC; push.(); IF pass => R = D).
        .
      "C := LDCT  := (Y = uPC; R = D).
        .
      "E := CONT  := (Y = uPC).
      end next

    end
```

comparisons provide an alternative way for evaluating the quality of the allocated designs. The design space for the AM2901 ISPS description is extensively explored.

11.1. Experiment on the AM2910

The AM2910 is a bipolar microprogram controller intended for use in high-speed microprocessor applications. Its data paths are depicted in Fig. 3. The controller contains a four-input multiplexer that is used to select either the register/counter (R), direct input (D), microprogram counter (uPC), or stack as the source of the next microinstruction address. The microprogram counter is composed of a 12-bit incrementer followed by a 12-bit register. Further details can be found in [2].

The VT file for the AM2910 contains 19438 bytes. Based on the VT, *Emerald* created a code sequence and allocated the data-memory. The CPU time for this run was 280 s on a VAX-11/780.

There are small differences between the code sequence and the original ISPS description. A segment of the original ISPS description is given in Table IX to illustrate the differences. As depicted in Table IX, the procedures *PUSH*, *LDCT*, and *CONT* contain the statement $Y = uPC$. *Emerald* combines the values for variables uPC and Y in the statement $Y = uPC$ of procedures *PUSH*, *LDCT*, and *CONT*. These operations are deleted. However, these data transfers are not eliminated. They appear as the data transfers for linking interface values between the calling and called VT-bodies. Therefore, they are moved into the context of the calling procedure Y.

Fig. 4 is the fastest implementation allocated by *Emerald*. Comparing Figs. 3 and 4, small differences can be observed. In Fig. 3 a wired-broadcast tree (*WBI*) is used to link D, R, and the multiplexer *MUXi*. Another wired-

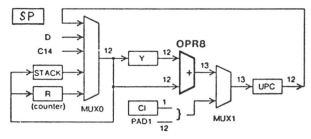

Fig. 4. Emerald's design for the AM2910.

TABLE X
QUANTITATIVE COMPARISONS FOR THE AM2910

Design Id	Emerald's Design	Commercial Design
# of Individual data operators	1	1
# of Gates Required for Data Operators	108	108
# of 9-bit Registers	1	1
# of 12-bit Registers	2	4
# of 13-bit Registers	2	0
Total No. of Bits	59	57
# of Gates Required for Storage Elements	472	456
Total No. of 2-input Multiplexers	61	36
# of Gates Required for Multiplexers	229	135
Total No. of Gates Required for the Design	809	699

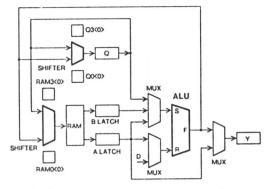

Fig. 5. Data paths of the commercial AM2901.

broadcast tree (*WB2*) is used to link the output port of *uPC*, input ports of *STACK*, and *MUXi*. To minimize the number of buses, *Emerald* merges these two wire-broadcast trees and *MUXi* into a larger multiplexer.

Emerald recognized that *R* is associated with the operation of incrementing by one and asked the designer to make a choice among counter, local incrementer, and ALU. In the commercial design *R* is implemented as a counter. For consistency, it was requested to be implemented as a counter.

In the commercial design an incrementer is used to perform the operation $uPC = Y + CI$. The "*CI*" is a single-bit register. To maintain consistency in size for operands of the operation, a padding operation is included in the value trace. A padding operation introduces a suitable number of zeros or ones to the most significant bit-positions of the original value. *Emerald* does not recognize that the operation can be implemented as a counter or incrementer. Instead, an ALU with a single add operation is allocated. The statement $uPC = Y + CI$ is translated into two statements: moving the output of the padding operation into *uPC* followed by $uPC = Y + uPC$.

C14 is a constant introduced in the ISPS description. This constant is not shown in the commercial design.

Table X compares the number of registers, data operators, and multiplexers used by the commercial and allocated designs. Since detailed information of the status bits is not available for the commercial design, comparisons are done only for the 12-bit data paths.

As indicated in the table, the allocated design uses two

more bits than the commercial design. These two bits are the *carry* bits of arithmetic operations. The number of data operators is the same for these two designs. As for interconnection units, the allocated design needs 61 while the commercial design uses only 36 two-input multiplexers. The value trace introduces one constant and one variable, which require additional interconnection units to direct the data flow. One constant (*C14*) increases the number of inputs of a multiplexer (*MUX0*) by one. The variable *CI* (*carry-in*) and the padding bits create an extra multiplexer (*MUX1*). The use of *C14* and *CI* accounts for 25 two-input multiplexers.

If the SN7483 is used as the adder or incrementer, the number of gates required for data operator is 108. Gate counts for other entities can be found in Table X. The total number of gates used by *Emerald's* design is 15.7 percent more than the commercial design.

The longest interconnection path for the commercial design consists of a four-input multiplexer (*MUXi*) while the longest interconnection path for Emerald's design consists of a five-input multiplexers. If the SN74153 and the SN74351 chips are used, these two data paths have the same delay of 17 ns.

11.2. Experiment on the AM2901

The AM2901 is a bit-sliced microprocessor chip which consists of a 16-word by 4-bit two-port RAM, a high-speed ALU, and the associated shifting, decoding and multiplexing circuitry. Its data paths are depicted in Fig. 5. Additional details can be found in [2].

The original ISPS description for the AM2901 contains eight procedures. Among them three procedures are used to compute the status bits. In practice, these status bits are not computed separately. They are the side effects of the arithmetic or logic operations and are, therefore, generated by a small number of AND, OR, and NOT gates. The three procedures for computing status bits introduce many data operations to the main data flow. To reduce the difficulty in analyzing the data, the algorithms for computing the status bits were deleted. Based on the remaining behavioral description, a VT file which contains 8421 bytes was generated. Using this VT file as input, the fastest implementation generated by *Emerald* is compared with the

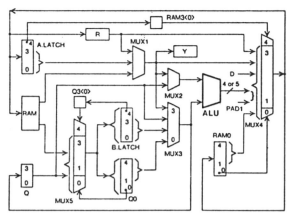

Fig. 6. Data paths allocated from the ISPS for the AM2901.

TABLE XI
A SEGMENT OF THE AM2901 ISPS DESCRIPTION

```
**Access Computation**{us}

source :=           ! Source calculation
 begin
 A.LATCH = RAM[A]; B.LATCH = RAM[B] next
 DECODE src =>
  begin
  #0 := (R = A.LATCH; S = Q
  #1 := (R = A.LATCH; S = B.LATCH),
  #2 := (R = 0      ; S = Q
  #3 := (R = 0      ; S = B.LATCH),
  #4 := (R = 0      ; S = A.LATCH),
  #5 := (R = D      ; S = A.LATCH),
  #6 := (R = D      ; S = Q
  #7 := (R = D      ; S = 0
  end
 end.

destination :=
 begin
 DECODE dest =>
  begin
  #0 := (Q = F; Y = F),
  #1 := (Y = F),
  #2 := (Y = RAM[A]; RAM[B] = F),
  #3 := (Y = F; RAM[B] = F),
  #4 := (Y = F; RAM[B] @ RAM0 = RAM3 @ F;
         Q @ Q0 = Q3 @ Q),
  #5 := (Y = F; RAM[B] @ RAM0 = RAM3 @ F),
  #6 := (Y = F; RAM3 @ RAM[B] = F @ RAM0;
         Q3 @ Q = Q @ Q0),
  #7 := (Y = F; RAM3 @ RAM[B] = F @ RAM0)
  end
 end.
```

TABLE XII
QUANTITATIVE COMPARISONS FOR THE AM2901

Design Id	Emerald's Design	Commercial Design
# of ALUs	1	1
# of Individual data operators	7	7
# of Gates Required for Data Operators	70	70
# of 1-bit Registers	2	4
# of 4-bit Registers	5	10
# of 5-bit Registers	5	0
# of 9-bit Registers	1	1
Total No. of Bits	56	53
# of Gates Required for Storage Elements	448	424
Total No. of 2-input Multiplexers	48	32
# of Gates Required for Multiplexers	180	120
Total No. of Gates Required for the Design	698	614

and F are assigned to the same register (represented as R in the allocated data paths). Finally, the statement $Q @ Q0 = Q3 @ Q$ are translated into two statements. The case becomes $Q0 \langle 4:0 \rangle = Q3 \langle 0:0 \rangle @ Q \langle 3:0 \rangle$ followed by $Q \langle 3:0 \rangle = Q0 \langle 4:1 \rangle$. A similar result was found for the shifting function associated with the input of the array memory.

Fig. 5 depicts the data paths of the commercial AM2901 while Fig. 6 illustrates the design produced by *Emerald*. Comparing the allocated design with commercial design leads to some observations. Since the ISPS description for the AM2901 contains simultaneous accesses to the *RAM*, a two-port memory was allocated. The memory data registers *A.LATCH*, *B.LATCH*, and *Y* are either connected to the first output port of the array memory through the multiplexer *MUX1* or connected to the second output port through the multiplexer *MUX5*. Another interesting observation is on the implementation of the shifting function. As indicated in Table XI, the ISPS description specifies the shifting function as a concatenation operation. The statement $Q @ Q0 = Q3 @ Q$ is an example. *Emerald* does not recognize this statement as a shifting operation. Instead, three registers $Q3$, Q, and $Q0$ which contain one, four, and five bits were allocated. These three registers and two multiplexers *MUX5* and *MUX3* are used to realize the shifting function. A similar structure was allocated for the shifting operation in front of the input port of the array memory. The associated data paths are *RAM0*, *RAM3*, *R*, *MUX1*, *MUX2*, and *MUX4*. Unlike Q, $Q0$, and $Q3$, R is not exclusively used for the shifting function. It is shared by other operations. If multiplexers for the shifting function are isolated from the rest of the data paths, the allocated data paths are quite similar to the commercial implementation.

Table XII compares the number of registers, data operators, and multiplexers used by the commercial and allocated designs. The following assumptions were made:

1) Y, which is a tri-state output buffer in the commercial design, is considered as a register.

commercial implementation. The CPU time for this run was 25 s on a VAX-11/780.

A segment of the original ISPS description for the AM2901 is given in Table XI. During the lifetime analysis, values generated by the variable Y are sometimes concluded as *dead values* (values which are defined and never used). Y was assumed to be a reserved variable. D represents an external input. It was intentionally put in the list of reserved variables in the beginning of the design process.

Comparing the resultant code sequence and the original ISPS description, some differences were observed. First, the variables S and Q are merged. The procedure *SOURCE* contains several statements that move the conent of Q into S ($S = Q$). These statements are eliminated. Another difference is that the values associated with the variables R

114

2) Each of the input and output ports of the ALU (*R*, *S*, and *F*) is counted as one register.
3) The external input (*D*) is considered as a register.

Including the two 4-bit memory address registers for the RAM, which are not shown in Fig. 5, the commercial design contains four 1-bit registers, ten 4-bit registers, and one 9-bit register. As indicated in the table, the allocated design uses 3 more bits than the commercial design. These bits are essentially allocated for the concatenation operations. The number of data operators is the same for these two designs. As for interconnection units, the commercial design uses twelve 3-input multiplexers[7] and eight 2-input multiplexers. This is equivalent to thirty-two 2-input multiplexers. Emerald's design needs 48 2-input multiplexers. Constants and registers for concatenation operations account for the necessity of the extra interconnection units.

If the SN74181 is used as the arithmetic and logic unit, the number of gates required for the data operator is 70. Gate counts for other entities can be found in Table XII. The total number of gates used by *Emerald's* design is 13.7 percent more than the commercial design.

The longest interconnection path for Emerald's design is from *A. LATCH* through *MUX*1 and *MUX*3, to the input of the ALU. This data path can be realized with two cascaded 4-input multiplexers. The longest interconnection path of the commercial design can be realized by a 4-input multiplexer. If the SN74153 chip is used, the typical delays of the commercial design and *Emerald's* design are 17 and 34 ns, respectively.

11.3. Exploring the Design Space for the AM2901 ISPS Description

A VT file which contains 16664 bytes was generated from the ISPS description of the AM2901. This ISPS description uses three procedures to compute condition codes and status bits. In these three procedures several operations can be simultaneously executed. *Emerald* creates designs by compaction. Therefore, the fastest implementation created by *Emerald* contains several ALU's. The significant difference in cost and speed between designs with different number of ALU's forms an interesting design space.

Looking at the initial code sequence, the maximum number of arithmetic and logic operations appearing in a *CodeList* is five. The design space is explored in such a way that the maximum number of arithmetic and logic operations in a *CodeList* is reduced by one at a time. Having serialized all the arithmetic and logic statements, simultaneous memory accesses were serialized. The last design resulted from serializing all the statements. In total, seven designs were generated. Data associated with these designs are summarized in Tables XIII–XVII[8]. The

---

[7]Each of the shifting functions contains four 3-input multiplexers.
[8]These data are for the AM2901 ISPS description which includes the three procedures for computing status bits. Table XII, on the other hand, contains data for the simplified ISPS description.

TABLE XIII
DATA ON STORAGE ELEMENTS

Design Id	#1	#2	#3	#4	#5	#6	#7
# of 1-bit Reg	13	13	13	13	13	13	13
# of 4-bit Reg	14	14	14	14	14	14	13
# of 5-bit Reg	5	5	5	5	5	5	5
# of 9-bit Reg	1	1	1	1	1	1	1
Total No. of Bits	103	103	103	103	103	103	99
# of Gates for Storage Elements	824	824	824	824	824	824	792

TABLE XIV
DATA ON OPERATORS

Design Id	#1	#2	#3	#4	#5	#6	#7
# of 1-bit ALUs	7	7	7	7	7	7	7
# of 1-bit Individual Data Operators	11	11	11	11	11	11	11
# of 4-bit ALUs	5	4	4	3	1	1	1
# of 4-bit Individual Data Operators	16	16	13	13	7	7	7
# of 3-bit ACUs	2	2	2	2	1	1	1
# of 3-bit Individual Comparators	3	3	3	3	2	2	2
# of 4-bit ACUs	1	1	1	1	1	1	1
# of 4-bit Individual Comparators	2	2	2	2	2	2	2
# of Gates for Data Operators	160	160	148	148	112	112	112

TABLE XV
DATA ON INTERCONNECTION UNITS

Design Id	#1	#2	#3	#4	#5	#6	#7
# of 2-input MUX	19	31	12	19	4	4	4
# of 3-input MUX	18	18	13	10	4	3	2
# of 4-input MUX	1	2	2	7	0	1	1
# of 5-input MUX	9	8	5	5	8	8	0
# of 6-input MUX	1	4	14	7	3	3	0
# of 7-input MUX	4	1	5	4	3	7	4
# of 8-input MUX	1	6	1	3	4	0	3
# of 9-input MUX	7	2	0	0	2	2	5
# of 10-input MUX	0	0	0	1	4	4	1
# of 11-input MUX	0	0	0	0	0	0	1
# of 13-input MUX	0	0	0	0	0	0	1
# of 14-input MUX	0	0	0	0	0	0	1
# of 16-input MUX	0	0	0	0	0	0	1
Equivalent Total No. of 2-input MUX	186	189	171	169	157	154	155
# of Gates for Multiplexers	698	709	642	634	589	578	582

calculation of the average number of steps required is based on three assumptions:

1) Equal probability is assumed for each branch associated with a branching operation.
2) The time for transfers of control is not considered.
3) Memory access is assumed to require one step.

Some observations on the data included in these tables are given below:

1) The structure of storage elements remains un-

TABLE XVI
COST AND PERFORMANCE FOR EACH DESIGN

Design Id	#1	#2	#3	#4	#5	#6	#7
Total No. of Gates Required	1682	1693	1614	1606	1525	1514	1486
Average No. of Steps	$17\frac{3}{4}$	18	$19\frac{1}{8}$	$20\frac{7}{8}$	24	25	$31\frac{5}{8}$

TABLE XVII
RUN TIME FOR EACH DESIGN

Unit: seconds

Design Id	#1	#2	#3	#4	#5	#6	#7
User Time	122.0	102.0	83.2	88.5	73.7	72.4	69.5
System Time	10.3	11.3	7.6	9.9	9.0	7.0	5.1
Total Run Time	132.3	113.3	90.8	98.4	82.7	79.4	74.6

changed for the first six designs. The code sequence for the last design was carefully arranged so that the live periods of some registers are separated. Therefore, the resulting allocation contains fewer bits than other designs. The association between values and allocated storage elements for these seven designs does show slight differences. Generally speaking, the tradeoffs on storage elements are not significant.

2) As expected, the number of aggregate and individual data operators strongly depends on the amount of parallelism. Serialization of arithmetic and logic operations generally reduces the number of aggregate and individual data operators.

3) Besides Design #2, the number of two-input multiplexers and the total number of gates required for a design decrease as the code sequence is serialized.

4) The effect of data paths on the performance is obvious. The ratio of speeds between the fastest and the slowest implementations is approximately 1.78.

5) The program was run on a Unix[9] operating system for the VAX-11/780. The run time for the initialization phase is approximately 76 seconds, which includes 62 seconds of user time and 14 seconds of system time.[10] The average run time for generating one design is slightly less than 96 s. The entire run for generating seven designs spends 748 seconds of CPU time (including run time for preparing the initial code sequence).

11.4. Experiment on the IBM System/370

The IBM System/370 is a mainframe computer. An ISPS description for its behavior is available on a computer system at Carnegie-Mellon University. The VT file translated from this ISPS description contains more than 1.5 Mbytes. The total number of input and output values for all the VT-bodies is more than 15 000. In order to handle large designs, Emerald was reimplemented with a checkpointing facility to allow partial processing to survive system crashes. Dubbed Emerald II, the new imple-

mentation successfully completed a design for the IBM System/370. It took 68 h to prepare the code sequence. Using the backup file for the code sequence as input, it took five and one-half hours to generate one design (30 min were spent in restoring the internal data structures). A design which took all the design options by default is presented in this section.

The ISPS description for the IBM System/370 contains specifications in register sets, constants, memory mapping, input/output operations, and instruction sets (including floating point instructions). Six array memories are declared: the main memory (MB), the storage keys (STKEYS), the general purpose registers (R), the floating point registers (FPREGISTER), and the floating error registers (FVU). The arithmetic and logic instructions essentially operate on 32-bit operands. However, 64-bit operations are included to perform multiplication, division, modulus, and dynamic address translation. In addition, floating point instructions introduce some 68-bit operations while memory mappings require 8-bit and 24-bit operations. Indeed, the IBM System/370 contains a large number of data types. The architectural details can be found in [16].

Comparing the code sequence generated by Emerald II and the original ISPS description, small deviations were observed. In ISPS the accesses to main memory may be of 8-bit, 16-bit, 32-bit, or 64-bit data transfers. In the VT these memory accesses are converted into a suitable number of 8-bit accesses. Another peculiarity is on the operations of reading and writing a subfield of a variable. In general, a constant is used to identify the offset of the subfield to be transferred. In the VT for the IBM System/370, some of these operations use a variable to specify the offset.[11] In this case the subfield to be transferred may change with time. To appropriately direct the data transfer, multiplexing and demultiplexing operators are introduced in the data paths. Besides these two points, the resultant code sequence stay fairly close to the original ISPS description.

The registers allocated by Emerald II can be classified into five categories. According to their function, they can be for global declarations, local declarations, procedure names with size attribute, formal parameters, and working storage such as temporary registers or status bits, etc.

As mentioned above, the IBM System/370 contains a large number of data types. The size of operations for these data types vary in a wide range. When the compatible graph for the data operators is constructed, Emerald checks the size of each pair of operators. If the difference in size is larger than a given constant, the pair is deleted from the graph. Emerald asks the designer to input the constant before allocating the aggregate data operators. The default value for the constant is one for ALU's and zero for ACU's. In other words, if the size of two arithmetic or logic operations differs in less than two bits, then they can be combined. Otherwise, they are separated. Two

---

[9]Unix is a trademark of AT&T Bell Laboratories.

[10]The system time is the time spent in serving operating system calls while the user time is the time used by the user program.

[11]For example, a variable is used to select the interrupt channel.

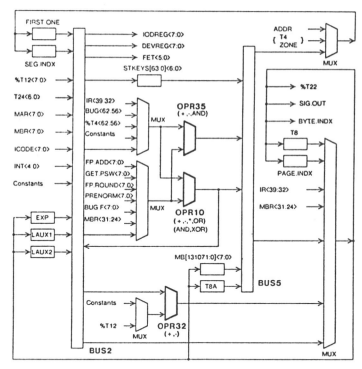

Fig. 7. 8-bit data paths of the IBM System/370.

TABLE XVIII
COMPONENT COUNTS FOR THE IBM SYSTEM/370

# of ALUs	26
# of Individual Data Operators in ALUs	64
# of ACUs	19
# of Individual Data Operators in ACUs	37
# of Multiplexing and Demultiplexing Operators	4
# of Gates Required for Data Operators	13950
# of Bits for Registers	2207
# of Bits for Shift Registers	522
Total No. of Bits	2729
# of Gates Required for Storage Elements	22224
Total No. of 2-input Multiplexers	7599
# of Gates Required for Multiplexers	28496
Total No. of Gates Required for the Design	64670

comparators can be combined only if they are of the same size. Therefore, many ALU's and ACU's were allocated and their sizes range from 1–112 bits.

Several interconnections result from the combination of global data flow and are, therefore, good candidates for a bus structure. Two of these structured interconnections which contain large number of input and output ports are represented as general buses. Data transfers between pairs of storage elements and data operators flow through these global buses.

The 8-bit data paths are depicted in Fig. 7. Component counts for various entities are given in Table XVIII. The total number of gates required for the design is estimated to be 64670. Further details of the design can be found in [25].

The longest interconnection path consists of a general bus followed by a 137-input multiplexer. If the SN74150 sixteen-input multiplexers are cascaded to form the 137-input multiplexer, typical delay for the longest data path is 46 ns.

XII. CONCLUSIONS

This paper presents a unifying procedure for the automated synthesis of data paths in digital systems. Based on the methodology, a design generator was implemented. Extensive experiments have been done with the design generator. Designs for the ISPS descriptions of the AM2910, the AM2901, and the IBM System/370 were selected to show the quality and limitations of automated synthesis. Generally speaking, if the design expert is restricted to the ISPS description, designs generated by *Emerald* are expected to be nearly identical to the design created by human experts. The design generator was able to generate data paths for a mainframe computer, the IBM System/370. Its practicality is verified.

The system can be enhanced in many ways such as augmenting the capability to recognize equivalent functions, providing the designer with the possibility of local modification, etc. Various scheduling algorithms for adjusting the code sequence can also be incorporated to facilitate different approaches of exploring a design space. Since the same procedure is applied to the allocation of storage elements, data operators, and interconnection units, investigating the effects of changing the allocation order for these three resources is also an interesting research. If the system is integrated into an automated design environment, the effort of digital design can be significantly reduced.

APPENDIX
THE CLIQUE-PARTITIONING ALGORITHMS

Let *G* be an undirected graph and its nodes be indexed by integers. Assume nodes *i* and *j* are connected and *i* is smaller than *j*. The edge which joins nodes *i* and *j* is represented by the integer-pair (*i*, *j*). Each of these nodes is the neighbor of the other node. If a third node is connected to the other two nodes, it is said to be a common neighbor of the pair. Two data structures are used to represent the graph. *NodeList* is used to store the nodes in the graph. It is a two-dimensional data structure. Each horizontal list contains a number of nodes which form a clique. Initially, each node of the graph occupies a horizontal list. When several nodes are grouped into a clique, they are coalesced into the same horizontal list. *EdgeList* is used to store the edges in the graph. All the edges which have the same "left node" (the node with the smaller index) are linked in a horizontal list. The indices in a horizontal list are sorted in an increasing order. The "left nodes" of all the horizontal lists are vertically linked together. Again, they are sorted in an increasing order.

Given a sorted list of edges of a graph, the following algorithm partitions the nodes of a graph into disjoint clusters. Each cluster forms a clique.

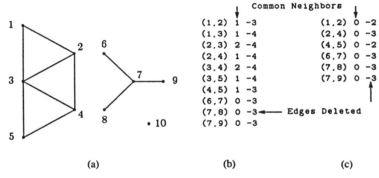

Fig. 8. Graph used by the example.

Algorithm 1: The Basic Clique-Partitioning Algorithm

1) Scan through the list of edges (*EdgeList*). For each (i, j), $i < j$, compute its number of common neighbors.

2) Pick the edge (p, q) which has the maximum number of common neighbors. Combine the lists of nodes headed by p and q. The smaller one of p and q is used as the head of the resulting clique.[12] Update the list of edges of G (as described in Algorithm 2). If the list of edges is empty, the graph partitioning is completed.

3) Assume p is the head of the resulting clique. Pick an edge which joins node p and other nodes and has the maximum number of common neighbors. Let the edge be (p, r), or (r, p) if r is smaller than p. Save r, or p if r is smaller than p. Update the list of edges of G (as described in Algorithm 2). If node p (or r if r is smaller than p) no longer appears in the *EdgeList*, go to Step 2 and start to collect the next cluster. Otherwise, repeat Step 3.

In Steps 2 and 3, if there is more than one pair having the maximum number of common neighbors, the numbers of edges which would be excluded are computed. The pair which excludes the least number of edges is selected. If more than one pair excludes the same number of edges, we choose one arbitrarily.

The number of common neighbors for each pair of connected nodes can be calculated by inspecting the list of edges. How is the number of edges to be excluded computed if a pair of nodes is grouped together? A node k which is connected to only one of i and j is no longer connected to the composite node (i, j). Thus the edge (i, k) or (j, k) must be deleted. A node k which is connected to both i and j is still connected to the composite node (i, j). Only one of edges (i, k) and (j, k) needs to be deleted. For consistency, each time one of these two edges needs to be deleted, the edge (j, k) is deleted. Therefore, the numbers of edges to be excluded can also be computed by inspecting the list of edges.

Once a pair of nodes is picked, the edge list needs to be updated in the following ways.

[12]Using the smaller node as the head of the resulting clique is just a matter of convenience. It does not influence the final result.

Algorithm 2: The Graph Updating Algorithm

1) Delete those edges which need to be deleted.
2) Recompute the numbers of common neighbors and the numbers of edges to be deleted for those pairs of connected nodes which remain in the list of edges.

An Illustrative Example

Let the graph depicted in Fig. 8(a) be given. The list of edges, the number of common neighbors and the edges to be deleted for each pair of connected nodes are depicted in Fig. 8(b). For example, the number of common neighbors and the number of edges to be excluded for (1, 2) can be computed in the following way. Node 3 is the only node which is connected to both nodes 1 and 2. The number of common neighbors for (1, 2) is thus one. If nodes 1 and 2 are grouped together, the edges (1, 2), (2, 3), and (2, 4) need to be deleted. Therefore, the number of edges to be excluded is three.

As indicated in Fig. 8(b), the pairs of nodes (2, 3) and (3, 4) have the maximum number of common neighbors and exclude the same number of edges if either pair is combined. Let nodes 2 and 3 be the first pair to be grouped together. Nodes 1 and 4 are connected to both nodes 2 and 3. They are connected to the composite node in the reduced graph. Node 5 is only connected to one of these two nodes, therefore, it is not connected to the composite node in the reduced graph. The list of edges of the reduced graph is depicted in Fig. 8(c).

To reduce the graph in Fig. 8(c), the numbers of common neighbors of the edges which consist of the composite node are compared. Both the edges (1, 2) and (2, 4) have the same number of common neighbors. Choosing the edge (1, 2), the number of edges to be excluded from the graph is less than choosing the edge (2, 4). Therefore, the edge (1, 2) is selected. The composite node which contains the nodes 1, 2, and 3 is no longer connected to other nodes. They belong to a cluster.

Repeatedly applying the procedure to the reduced graph, the nodes in the original graph are partitioned into six clusters. They are {1, 2, 3}, {4, 5}, {6, 7}, {8}, {9}, and {10}.

As mentioned in Section IV, direct application of the basic clique-partitioning algorithm to the data-memory allocation problem does not generate good solutions. And,

IEEE TRANSACTIONS ON COMPUTER-AIDED DESIGN, VOL. CAD-5, NO. 3, JULY 1986

the notion of divide-and-conquer as well as the notion of transitive property are introduced to overcome the problem of randomly selecting nodes to merge. The following paragraphs present the generalized clique-partitioning algorithm.

Let (p, q), (p, r) [or (r, p) if r is smaller than p], and (q, r) [or (r, q) if r is smaller than q] be three edges in G where p is smaller than q. As indicated in Algorithm 1, if p and q are combined, then the composite node is represented by the smaller node p. Furthermore, the edge (q, r) is deleted from G.

Let (p, r), (q, r), and (p, q) belong to three different categories, which are represented by i, j, and k. Assume the profits of grouping a pair of nodes in categories i, j, and k are ordered in an increasing manner. In addition, these three edges have the following form of transitive property (named the generalized transitive property). If the nodes p and q are grouped together, then nodes p and r can be included in a new category l, where l is the lower case of L. The edges in category l have a profit measure better than edges in category i. The algorithm is given below.

Algorithm 3: A Generalized Clique-Partitioning Algorithm

1) Scan through the list of edges of category $k(Gk)$. For each (i, j), compute its number of common neighbors.
2) Pick the edge (p, q) which has the maximum number of common neighbors from Gk.
3) Instead of directly applying Algorithm 2 to G and Gk, update G and Gk in the following way. For each node r which is only connected to one of the nodes p and q in the graph G, the edge is deleted from G. If the edge is also an edge of category k, it is deleted from Gk. For each node r which is connected to both p and q in G, the edge (q, r) is deleted from G. If this edge is contained in Gk, it is also deleted. Assume that (p, r) is an edge of category i and (q, r) is in the list of edges for category j. Due to the combination or grouping of p and q, the edge (p, r) becomes an edge of category l. The category indentifier of (p, r) is changed to l. If the profit of combining pairs of nodes in category l is the same as or better than that of combining pairs of nodes in category k, the edge is included in Gk. Meanwhile, the number of common neighbors and the number of edges to be excluded for each of the edges remaining in Gk are updated.

The subgraph in which pairs of nodes have the best profit measure is reduced first. Then the pairs of nodes having the next level of profit measure are collected and reduced. Repeatedly applying the procedure to the other subgraphs, the process is stopped when a subgraph of a specified category or the original graph G becomes empty.

The transitive properties defined in Sections V–VII assume that the category identifiers j, k, and l refer to the same category. This is actually a special case of the generalized transitive property. It is named the loose form transitive property.

ACKNOWLEDGMENT

Dr. Stephen W. Director, Dr. Dario Giuse, and Dr. Donald E. Thomas have provided many constructive remarks. Special thanks are due to Bill Birmingham for reading and polishing early drafts of this paper. Finally, the authors are grateful to the referees for many suggestions which have improved the clarity of the presentation.

REFERENCES

[1] A. V. Aho and J. D. Ullman. *Principles of Compiler Design.* Reading, MA: Addison-Wesley, 1977.
[2] *The AM2900 Family Data Book.* Advanced Micro Devices, Inc. 1976.
[3] J. G. Augustson and J. Minker, "An analysis of some graph theoretical cluster techniques," *J. ACM* 17(4):571–588, Oct. 1970.
[4] M. R. Barbacci, "Automated Exploration of the Design Space for Register Transfer (RT) systems," Ph.D. dissertation, Carnegie-Mellon Univ., Nov. 1973.
[5] M. R. Barbacci, G. E. Barnes, R. G. Cattell, and D. P. Siewiorek, "The symbolic manipulation of computer descriptions: ISPS computer description language," Carnegie-Mellon Univ., 1979.
[6] C. Bron and J. Kerbosch. "Finding all cliques of an undirected graph—Algorithm 457," *Commun. ACM*, 16(9):575–577, September, 1973.
[7] J. Darringer and W. Joyner. "A new look at logic synthesis," in *17th IEEE Design Automation Conf. Proc.* p. 543–549, June 1980.
[8] S. W. Director, A. C. Parker, D. P. Siewiorek, and D. E. Thomas, "A design methodology and computer aids for digital VLSI systems," *IEEE Trans. Circuits Syst.*, vol. CAS-28, p. 634–645, July 1981.
[9] L. J. Hafer and A. C. Parker, "Automated synthesis of digital hardware," *IEEE Trans. Computers*, vol. C-31, pp. 93–109, Feb. 1982.
[10] L. J. Hafer and A. C. Parker, "A formal method for the specification, analysis, and design of register–transfer level digital logic," *IEEE Trans. Computer-Aided Design*, vol. CAD-2, pp. 4–18, Jan. 1983.
[11] P. Marwedel, "The MIMOLA design system: Detailed description of the software system," in *16th Design Automation Conf. Proc.*, pp. 59–63, 1979.
[12] G. McClain, "Optimal design of central processor data paths," Ph.D. dissertation, Dep. Electrical and Computer Eng., University of Michigan, Apr. 1972.
[13] M. C. McFarland, "The VT: A database for automated digital design," Technical Report DRC-01-4-80, Design Research Center, Carnegie-Mellon University, December, 1978.
[14] G. D. Mulligan and D. G. Corneil, "Corrections to Bierstone's Algorithm for Generating Cliques," *J. ACM* 19(2):244–247, Apr. 1972.
[15] M. C. Paull and S. H. Unger, "Minimizing the number of states in incompletely specified sequential functions," *IRE Trans. Electronic Computers*, vol. EC-8, pp. 356–367, Sept. 1959.
[16] N. S. Prasad, "Architecture and Implementation of Large Scale IBM Computer Systems," Q.E.D. Information Sciences, Inc., Wellesley, MA, 1981.
[17] S. L. Rege, "Designing variable data format modules with cost-performance tradeoffs," Ph.D. dissertation, Carnegie-Mellon Univ., Aug. 1974.
[18] E. M. Reingold, J. Nievergelt, and N. Deo, *Combinatorial Algorithms: Theory and Practice.* Englewood Cliffs, NJ: Prentice-Hall 1977.
[19] H. E. Shrobe, "The data path generator," in *Proc. Conf. Advanced Research in VLSI*, MIT, Cambridge, MA, Jan. 1982, pp. 175–181.
[20] E. A. Snow, "Automation of module set independent register–transfer level design," Ph.D. dissertation, Carnegie-Mellon University, Apr. 1978.
[21] *The TTL Data Book for Design Engineers.* Texas Instruments, Inc., 1981.
[22] D. E. Thomas, C. Y. Hitchcock III, T. J. Kowalski, J. V. Rajan, and R. A. Walker, "Automatic data path synthesis," *IEEE Computer*, vol. 16, pp. 59–73, Dec. 1983.
[23] C. J. Tseng and D. P. Siewiorek, "The modeling and synthesis of bus systems," in *Proc. Eighteenth Design Automation Conf.*, pp. 471–478. June 1981.

[24] C. J. Tseng and D. P. Siewiorek, "A note on the automated synthesis of bus style systems," Tech. Rep. Dep. of Electrical Eng., Carnegie-Mellon University, October, 1982.

[25] C. J. Tseng, "Automated synthesis of data paths in digital systems," Ph.D. dissertation, Dep. of Electrical and Computer Engineering, Carnegie-Mellon University, Pittsburgh, PA, Apr. 1984.

[26] S. Tsukiyama, M. Ide, H. Ariyoshi, and I. Shirakawa, "A new algorithm for generating all the maximal independent sets," *SIAM J. Computing*, vol. 6, no. 3, pp. 505–517, Sept. 1977.

[27] G. Zimmerman, "The MIMOLA design system: A computer aided digital processor design method," in *16th Design Automation Conf. Proc.*, pp. 53–58, 1979.

A Formal Method for the Specification, Analysis, and Design of Register–Transfer Level Digital Logic

LOUIS J. HAFER, MEMBER, IEEE, AND ALICE C. PARKER, MEMBER, IEEE

Abstract—This paper describes a method for formally modeling digital logic using algebraic relations. The relations model digital logic at the register–transfer (RT) level. An RT-level behaviorial specification is used to develop the relations, which express timing relationships that must be satisfied by any correct implementation. An extension of the model is shown which can be used for synthesis at the RT level. The growth rate and computational properties of the model are discussed, and an example of synthesis is shown.

Manuscript received January 4, 1982; revised August 24, 1982. This work was supported in part by the National Science Foundation under Grant ENG 78-25755, and in part by the IBM corporation under a fellowship.

L. J. Hafer is with the Department of Computing Science, Simon Fraser University, Burnaby, British Columbia, Canada V5A 1S6.

A. C. Parker is with the Department of Electrical Engineering Systems, University of Southern California, Los Angeles, CA 90007.

I. INTRODUCTION

IN GENERAL, the problem faced by digital designers is to create a hardware implementation of a digital system which exhibits a specified behavior, satisfies any constraints imposed on it, and is "optimal" with respect to some set of design goals. Often, the design goals compete (cost and performance, for ex-

Reprinted from *IEEE Trans. CAD of Int. Circ. Syst.*, vol. CAD-2, no. 1, pp. 4–18, Jan. 1983.

ample), and in this case the best implementation will be the one that embodies the most acceptable tradeoff between the various objectives.

Computer-aided design/design automation can aid a designer in selecting the best implementation by allowing him to generate alternative designs faster than an unaided designer, permitting a greater range of designs to be evaluated. If the designer is to receive the maximum benefit from this ability, the designs which are generated must accurately reflect the tradeoffs between the design goals. They must be *noninferior* designs; i.e., designs where increasing the degree of satisfaction of any one objective can only be accomplished at the expense of other objectives. The use of an algebraic model to express the design problem allows the power of formal constrained optimization techniques to be brought to bear on this aspect of the synthesis problem.

At the register–transfer (RT) level, design is not a formalized process. There is no technique analogous to circuit analysis or Boolean algebra which is applied at the RT level. The design specifications are often provided in a mixture of forms (e.g., timing diagrams, flowcharts, text), and the implementation is constructed using a collection of independent rules and procedures which are applied on a case-by-case basis until the designer is satisfied that the design constraints have been met. The performance and correctness of the design are often determined by simulation, and subsequent design changes require reevaluation by the same means. Many methods for applying this collection of rules and procedures have been proposed, but no one has publicly announced a unifying representation for them.

Consider an example where two distinct operations are to be performed on the same two operands. Assume there is no *a priori* ordering between the two operations, and that both results must be available concurrently for some period of time. An example RT level design rule applicable here is "the two results may be obtained in parallel using one operator for each operation, or serially, using a single operator capable of both operations and a storage element to hold the first result."

This rule allows a designer to optimize the operation sequence for performance. The usefulness and desirability of such an optimization may, however, be difficult to ascertain. The relative costs and speeds of the hardware elements must be taken into account. In the general case, where the example operations are part of a larger behavioral description, the surrounding context must also be considered. Some hardware elements may be required to implement other operations. Also, there may be operation sequences executing in parallel with the example operations which require more time than either alternative suggested above. Finally, subsequent optimizations may change any or all of these factors.

Two general observations can be made about the RT level design process. First, the rules and procedures which are used relate specifically to resource utilization, proper ordering of operations, and timing relationships. Second, the interactions between the design constraints, the hardware elements used in the implementation, and the characteristics of the finished implementation are complex, and difficult to express in words.

This paper asserts that 1) both the behavioral specification and the RT level design rules can be modeled by a system of algebraic relations, and 2) simultaneous solution of this system of algebraic relations as a constrained optimization problem allows global optimization of the design, since all interactions between the components of the design problem are accounted for in the algebraic relations.

The remainder of the paper supports these assertions by proposing a method for expressing, with algebraic relations, the conditions which must be satisfied by a correct RT level design. The relations encompass the behavior the design must support and the performance constraints it must satisfy. These relations are then extended to include RT level design choices, and a method for generating the relations automatically from the behavioral specification is outlined. When this system of relations is linearized and solved as a mixed-integer linear programming (MILP) problem, the solution yields both a design for the data part and a timing specification which details how the behavioral actions are to be executed on the data part.[1] The design is optimal with respect to the objective function supplied by the designer for evaluating candidate implementations. Optimal results can be easily obtained for small designs or local modifications to existing designs. Unfortunately, computational complexity appears to limit the size of designs for which unspecialized MILP is a feasible solution technique, but the use of specialized algorithms offers hope for relaxing this limitation.

There are several advantages to this method for modeling digital design. The primary advantage of a formal model is the insight it provides into the nature of the digital design problem. By modeling an existing implementation using algebraic relations, the many relationships that exist between the hardware elements are explicitly revealed. Similarly, when a model is created for synthesis, the effects of design decisions are also made explicit, allowing the designer to clearly see the effects of each decision. Heuristic technqiues can be developed and evaluated by examining their power and accuracy using a formal model.

The methodology also has practical advantages. The ease with which portions of a design can be "frozen" while other portions are changed makes the method attractive for iterative design and engineering changes. It also makes it easy for the designer to intervene to place arbitrary constraints on the design. Verification of existing designs can also be done using this modeling method, without the computational problems mentioned above.

A. Related Research

For the interested reader, the following references can provide information on other work related to the subject of this paper.

Recent research in computer-aided design and design automation for logic synthesis using a heuristic approach includes the CMU-DA system, the MIMOLA system, and work by Darringer *et al.* The CMU-DA system [1], [8], [11], [16],

---

[1] By data part, the authors mean the hardware elements, both storage and operator, and the interconnections which are used to transfer, store, and transform data values.

[17], [20] is the product of an ongoing research effort at Carnegie–Mellon University which applies a hierarchical design philosophy to the problem of creating a hardware implementation from an initial behavioral specification. The design process is one of binding implementation decisions in a top-down manner as a design proceeds through the system. More and more structural detail is frozen at each level until a complete hardware specification is obtained.

The MIMOLA system [13], [21] acts as a hardware compiler, using complex hardware modules. Starting from a behavioral description that emphasizes parallelism in the algorithm, the user guides the system by restricting the set of available hardware modules.[2] Darringer et al. [5] have concentrated on generating master-slice implementations from structural RT level descriptions. Their approach emphasizes the use of local, computationally efficient transforms to generate an acceptable implementation.

Other researchers have used formal, algebraic descriptions to define the problem of generating an implementation. Work by Hinshaw [10] and by McClain [14] falls into this category. The techniques used by both tend to cast the synthesis problem into the form of an optimal set selection problem. Simplifying assumptions are made about the sequence in which operations in the behavioral description are executed on the allocated hardware, and an optimal set of hardware modules is selected.

Analogous research also exists in other problem domains. Hachtel [7] has proposed a mixed-integer linear programming formulation of the placement problem for further investigation. Chu [3] has used 0/1 integer programming for task allocation in a distributed processing environment, and Ciesielski [4] has used the same formalism to minimize vias in multilayer routing problems. Finally, Grossmann and Papoulias [6], [18] have used mixed-integer linear programming models for the design of chemical plants.

II. DEVELOPING A FORMAL MODEL

This section presents a model which specifies the basic timing relations that must be satisfied by an RT level data part implementation, if the implementation is to support the actions of a RT level behavioral specification. For clarity and continuity of exposition, this model makes assumptions about the generic characteristics of the hardware elements modeled, and about the level of detail included in the model. The complexity of a model increases with the amount of detail included; this development aims for a level sufficiently detailed to illustrate the power of the modeling methodology without obscuring the basic concepts. It must be emphasized that the model developed here is a specific example of the kinds of models which can be developed within the general methodology of expressing timing and structural properties as algebraic relations. Examples of other, more detailed models can be found in the original thesis [9].

The basic feature modeled here is the transfer of data through one or more operators to a storage element. The model will be

---

[2] The design is iteratively improved until the hardware resources are used efficiently.

Fig. 1. (a) Examples ISPS behavioral description: calc. (b) Data flow representation of calc.

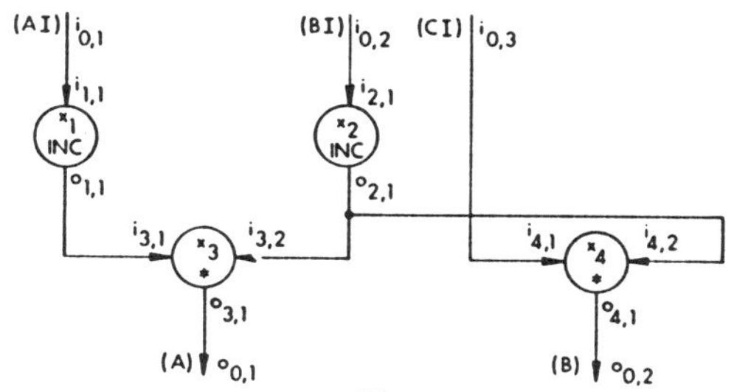

concerned with 1) the proper ordering of operations and data stores to ensure correct behavior, and 2) the timing involved in carrying out each transfer. Data paths and switching logic are not explicitly modeled; in essence, point-to-point data paths with no delay are assumed. The hardware elements used in example implementations are assumed to have generic characteristics patterned after TTL SSI/MSI IC's, with two restrictions. Operation and storage capabilities are not available in a single unit, thus excluding devices such as shift registers and counters, and actions are assumed to be evoked by rising or falling edges, rather than by signal levels.

A. Representations and Definitions

The algebraic relations that make up the model are derived from an augmented data flow representation that expresses the original RT level behavioral specification in terms of operations and values. The "augmented" quality stems from the explicit representation of the control flow effects as well as the associated data flow of control constructs. This paper will use an informal version of the Value Trace representation implemented by McFarland [15].

To make this idea of a data flow representation more concrete consider the example shown in Fig. 1. Fig. 1(a) is a simple two line ISPS behavioral description specifying two data transfers. The "next" indicates that the designer has assigned an ordering in time to the two assignment statements. This ordering is arbitrary and does not appear in the corresponding data flow. Assuming that the initial values of AI, BI, and CI are obtained from the external environment and that the final values of A and B must be passed back to that environment, the data flow representation of Fig. 1(a) is shown in Fig. 1(b). Note that the act of calculating $(BI + 1)$ does not change the value of BI; this is reflected in the data flow representation by distributing the resulting value from a single operation node.

The variable names in parentheses indicate the correspondence between the value arcs of the data flow graph and the ISPS variables. The nodes labeled x_1 and x_2 correspond to the addition (INCrement) operations in the ISPS, and those labeled x_3 and x_4 to the multiply operations.

A few sentences about the use of the words "operation" and "operator" are in order here, as they have specific meanings in this paper. *Operation* will always mean an abstract operation in the behavioral description, and will most often

TABLE I
MNEMONICS USED FOR VARIABLE NAMES

	started/stored	available	released
operation input		$T_{IA}(i_{a,b})$	$T_{IR}(I_a)$
operation output		$T_{OA}(O_a)$	$T_{OR}(o_{a,c})$
operation	$T_{XS}(x_a)$		$T_{XR}(x_a)$
stored copy of operation output	$T_{SS}(o_{a,c})$	$T_{SA}(o_{a,c})$	$T_{SR}(o_{a,c})$

TABLE II
TIMING VARIABLES FOR THE BASIC MODEL

$T_{IA}(i_{a,b})$	Time when the value required by input $i_{a,b}$ of operation x_a is available for use in the computation.
$T_{XS}(x_a)$	Time when the computation of operation x_a actually begins.
$T_{OA}(O_a)$	Time when the output values O_a computed by operation x_a are available at the outputs of the operator performing the operation.
$T_{XR}(x_a)$	Time when the output values O_a of operation x_a are no longer required as input values for another operation, and thus the time that execution of the operation can cease.
$T_{IR}(I_a)$	Time when the input values for operation x_a are no longer required.
$T_{OR}(o_{a,c})$	Time when output value $o_{a,c}$ is no longer required as an input value for another operation.
$T_{SS}(o_{a,c})$	Time when the storage element assigned to store output value $o_{a,c}$ is clocked
$T_{SA}(o_{a,c})$	Time when the value $o_{a,c}$ is available at the output of the storage element assigned to store it.
$T_{SR}(o_{a,c})$	Time when the stored copy of value $o_{a,c}$ is no longer required as an input value for an operation.

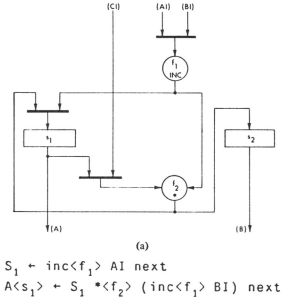

(a)

$$S_1 \leftarrow inc\langle f_1 \rangle\ AI\ next$$
$$A\langle s_1 \rangle \leftarrow S_1\ *\langle f_2 \rangle\ (inc\langle f_1 \rangle\ BI)\ next$$
$$B\langle s_2 \rangle \leftarrow CI\ *\langle f_2 \rangle\ (inc\langle f_1 \rangle\ BI)$$

(b)

Fig. 2. (a) Example data part for the calc ISPS. (b) Structural RT level operation sequence for the example data part.

be used in the context of the data flow representation. *Operator* will always mean a hardware element used to implement an operation, and will always be used in the context of a hardware implementation.

The input and output ends of value arcs are labeled in a like manner. The inputs to operation x_a are labeled $i_{a,b}$, and the outputs $o_{a,c}$, where the subscripts b and c distinguish individual values.

Fig. 1(b) also illustrates the naming conventions for components of the data flow representation. Operation nodes are labeled x_1, x_2, etc., and x_a indicates an operation in general. o_a indicates all outputs from operation a; $o_{a,c}$ indicates the oth output. By convention, a 0 (zero) operation subscript indicates the external environment, so $i_{0,c}$ is an input from the external environment, and $o_{0,b}$ is an output to the external environment. The apparent inconsistency in the use of the b and c subscripts can be resolved by considering the external environment as one complex activity with inputs $o_{0,b}$ and outputs $i_{0,c}$. This view corresponds to the way these values are actually handled in the model.

In order to write algebraic relations describing the proper timing and sequencing properties required for correct behavior, variables must be defined to represent certain critical times. These are associated with the operations and values of the data flow specification, and are defined in Table II. The mnemonic scheme used to name the variables is shown in Table I. Some of the variables are redundant (i.e., defined by equalities from other variables) but are included here for clarity.

Fig. 2 shows a possible implementation of a data part or the behavior specified in Fig. 1. Fig. 2(a) shows the data part hardware, and Fig. 2(b) gives a structural RT level description

TABLE III
HARDWARE TIMING VALUES

$D_{FP}(f_d)$	Propagation delay time of operator f_d from the appearance of the input value(s) at the operator input(s) to the appearance of the output value(s) at the operator output(s).
$D_{SS}(s_e)$	Setup time at the data input of storage element s_e: data at the input to storage element s_e must be valid for at least this long prior to a transition at the clock input of the storage element.
$D_{SH}(s_e)$	Hold time at the data input of storage element s_e: data at the input of storage element s_e must remain valid for at least this long after the transition at the clock input of the storage element.
$D_{SP}(s_e)$	Propagation delay time of storage element s_e from the transition at the clock input to the appearance of the value at the storage element output

of the actions necessary to execute the given behavior using this data part. As illustrated in Fig. 2(a), the labeling convention for hardware elements is that operators are labeled f_d (e.g., f_1, f_2) and storage elements are labeled s_e (e.g., s_1, s_2). The hardware labels in Fig. 2(b) indicate the hardware element associated with a particular action. The heavy horizontal bars in Fig. 2(a) indicate switching elements, and are not labeled since the simplified model presented here will not deal explicitly with data paths and switching. Table III shows the set of time delays that are associated with hardware elements. For any specific element, these values are assumed to be available from data sheets or similar sources.

In the development which follows, Figs. 1 and 2 will be used to provide specific examples. The development will be divided into three main segments: relations which model the execution of an operation on an operator, relations which model storing a value in a storage element, and relations which ensure that hardware resources are shared correctly. Figs. 3 and 4 are graphical representations of the algebraic relations presented in Sections II-B and II-C, respectively.

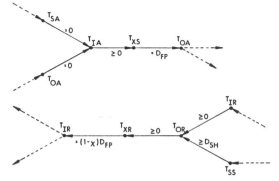

Fig. 3. Timing relations for operation execution.

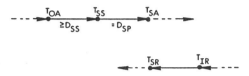

Fig. 4. Timing relations for storing a value.

B. Modeling Operations

Before an operation can be executed, all the necessary input values must be present at the inputs of the operator executing the operation. $T_{IA}(i_{a,b})$ is the time the value required at input $i_{a,b}$ will be available, and it must satisfy one of two relations. If the value is to be obtained from a stored copy, as is the case for input $i_{3,1}$ in the example, then it must be that

$$T_{IA}(i_{3,1}) = T_{SA}(o_{1,1}).$$

Simply, the time $T_{IA}(i_{3,1})$ when the value $o_{1,1}$ becomes available at the operator input corresponding to input $i_{3,1}$ will be the time $T_{SA}(o_{1,1})$ when the value is available at the output of the storage element. (Recall that, for clarity, this model assumes data paths and switching logic do not introduce propagation delays.) On the other hand, if the value is to be obtained from an operator output, as is the case for input $i_{3,2}$, it must be that

$$T_{IA}(i_{3,2}) = T_{OA}(O_2).$$

In general, if input $i_{a2,b2}$ requires value $o_{a1,c1}$, the time $T_{IA}(i_{a2,b2})$ when the value is available must satisfy either

$$T_{IA}(i_{a2,b2}) = T_{SA}(o_{a1,c1}) \tag{1}$$

or

$$T_{IA}(i_{a2,b2}) = T_{OA}(O_{a1}). \tag{2}$$

Continuing to use operation x_3 as an example, the time $T_{XS}(x_3)$ when operation x_3 begins to execute using operator f_2 must satisfy the inequalities

$$T_{XS}(x_3) \geqslant T_{IA}(i_{3,2})$$

and

$$T_{XS}(x_3) \geqslant T_{IA}(i_{3,1}).$$

The inequalities simply state that operation x_3 cannot be started until both input values are available at the operator

inputs. In general, the relation is

$$T_{XS}(x_a) \geqslant T_{IA}(i_{a,b}) \tag{3}$$

and this inequality must be satisfied for all inputs $i_{a,b}$ of operation x_a.

Once an operation starts, a time equal to the propagation delay of the operator performing the operation must elapse before the output is available. Operation x_3 executes on operator f_2, and the time $T_{OA}(O_3)$ when the output is available is given by the equation

$$T_{OA}(O_3) = T_{XS}(x_3) + D_{FP}(f_2).$$

In general, if operator f_d implements operation x_a, it must be that

$$T_{OA}(O_a) = T_{XS}(x_a) + D_{FP}(f_d). \tag{4}$$

The relations above model the forward aspect of an operation, i.e., the initiation of the operation to produce output values from input values. To be complete, relations are needed to model the backward aspect, i.e., the lifetime of the outputs and the cessation of the operation when the output values are no longer needed.

Each output value must remain valid until it is no longer needed by the inputs that use the value. There are two cases to consider, as a value may be stored, or used directly as the input to another operation. Consider the output value $o_{1,1}$, which is stored in storage element s_1. The lifetime $T_{OR}(o_{1,1})$ of the output must satisfy

$$T_{OR}(o_{1,1}) \geqslant T_{SS}(o_{1,1}) + D_{SH}(s_1).$$

This inequality states that output $o_{1,1}$ must remain valid for at least the hold time $D_{SH}(s_1)$ after the time $T_{SS}(o_{1,1})$ when the register is clocked to latch the value. Output of $o_{2,1}$, on the other hand, is accessed directly by operation inputs $i_{3,2}$ and $i_{4,2}$. Thus $T_{OR}(o_{2,1})$ must satisfy the relations

$$T_{OR}(o_{2,1}) \geqslant T_{IR}(l_3)$$

and

$$T_{OR}(o_{2,1}) \geqslant T_{IR}(l_4).$$

In general, an output $o_{a1,c1}$ could be both stored and used by other operations, so that its lifetime $T_{OR}(o_{a1,c1})$ must satisfy

$$T_{OR}(o_{a1,c1}) \geqslant T_{SS}(o_{a1,c1}) + D_{SH}(s_e) \tag{5}$$

if the value is stored in storage element s_e, and

$$T_{OR}(o_{a1,c1}) \geqslant T_{IR}(l_{a2}) \tag{6}$$

for every operation input $i_{a2,b2}$ which uses it directly.

Given the lifetimes of the outputs of an operation, it is possible to construct relations for the time that the operation can cease and the time the operation inputs are no longer needed. Operation x_2 has one output, $o_{2,1}$, so the operation can cease only after that output is no longer required as an input to another operation. This is expressed by

$$T_{XR}(x_2) \geqslant T_{OR}(o_{2,1}).$$

In the general case of an operation x_a, $T_{XR}(x_a)$ must satisfy

$$T_{XR}(x_a) \geqslant T_{OR}(o_{a,c}) \tag{7}$$

for all outputs $o_{a,c}$ of the operation.

When an operation ceases, its inputs are no longer needed. Indeed, they can be released slightly earlier, since the outputs will not change until the propagation delay has elapsed after the input change. For operation x_2, executed on operator f_1, the time the inputs can be released is

$$T_{IR}(l_2) = T_{XR}(x_2) - (1 - \chi)D_{FP}(f_1), \quad 0 \leqslant \chi \leqslant 1.$$

χ can be thought of as a "safety factor", allowing a designer to delay the release of the inputs, rather than taking full advantage of the operator propagation delay. In general, for inputs l_a of operation x_a, performed on operator f_d, it must be that

$$T_{IR}(l_a) = T_{XR}(x_a) - (1 - \chi)D_{FP}(f_d), \quad 0 \leqslant \chi \leqslant 1. \tag{8}$$

This equation completes the set of relations needed to model an operation.

C. Modeling the Store of a Value

In order to correctly store a value, the setup time of the storage element must be observed. To store the output value $o_{1,1}$ in storage element s_1, for example, the clock time $T_{SS}(o_{1,1})$ must satisfy the relation

$$T_{SS}(o_{1,1}) \geqslant T_{OA}(O_1) + D_{SS}(s_1).$$

In the general case, the relation becomes

$$T_{SS}(o_{a,c}) \geqslant T_{OA}(O_a) + D_{SS}(s_e). \tag{9}$$

Just as with operators, a propagation delay must elapse before the storage element output is valid. For $o_{1,1}$, the equation would be

$$T_{SA}(o_{1,1}) = T_{SS}(o_{1,1}) + D_{SP}(s_1)$$

and in general the equation is

$$T_{SA}(o_{a,c}) = T_{OS}(o_{a,c}) + D_{SP}(s_e). \tag{10}$$

Finally, a relation is needed for the lifetime of the value in the storage element. Again using output value $o_{1,1}$ as the example, the lifetime $T_{SR}(o_{1,1})$ of the value in the storage element must satisfy the inequality

$$T_{SR}(o_{1,1}) \geqslant T_{IR}(l_3)$$

since the stored copy of the value is used by input $i_{3,1}$. Since this model does not consider the direct transfer of a value from one storage element to another, no second relation analogous to (5) is presented. In general, the relation

$$T_{SR}(o_{a1,c1}) \geqslant T_{IR}(l_{a2}) \tag{11}$$

must be satisfied for all input sets l_{a2} for which some input $i_{a2,b2} \in l_{a2}$ uses the stored value.

This completes the set of relations required to model the storage of a value.

D. Modeling Hardware Resource Sharing

In most RT level designs, storage elements and operators are used frequently to store several different values or perform several operations (in the sense of the operations in the data flow representation). Relations must be constructed which will ensure that the same hardware resource is not scheduled to perform more than one function during any given time interval.

Consider two closed intervals of time, $[t1, t2]$ and $[t3, t4]$, where $t1 < t2$ and $t3 < t4$. We will define an overlap function L to be

$$L([t1, t2], [t3, t4]) = \begin{cases} 1, & \text{if the intervals overlap} \\ 0, & \text{otherwise.} \end{cases}$$

Now, referring to Figs. 1 and 2, the example shows that both multiply operations, x_3 and x_4, are executed by the operator f_2. The constraints of Sections II-B and II-C are basically mathematical expressions of the data dependencies in the data flow representation.[3] Fig. 1(b) shows that no such dependencies exist between the operations x_3 and x_4. Using operation x_3 as an example, execution begins on operator f_2 at the time $T_{XS}(x_3)$, and ceases at the time $T_{XR}(x_3)$. Thus the closed interval $[T_{XS}(x_3), T_{XR}(x_3)]$ represents the time interval when operator f_2 is occupied with executing operation x_3. Similarly, the interval $[T_{XS}(x_4), T_{XR}(x_4)]$ gives the time interval when operator f_2 is occupied with operation x_4. To ensure that they are not both scheduled to be executed at the same time, the relation

$$L([T_{XS}(x_3), T_{XR}(x_3)], [T_{XS}(x_4), T_{XR}(x_4)]) = 0$$

must be satisfied so that execution intervals do not overlap. In the general case, the relation

$$L([T_{XS}(x_{a1}), T_{XR}(x_{a1})], [T_{XS}(x_{a2}), T_{XR}(x_{a2})]) = 0 \tag{12}$$

must be satisfied for all pairs of operations x_{a1} and x_{a2} which use the same operator f_d. For storage elements, the time interval during which the storage element is occupied by a value $o_{a,c}$ is $[T_{SS}(o_{a,c}), T_{SR}(o_{a,c})]$, and the relation

$$L([T_{SS}(o_{a1,c1}), T_{SR}(o_{a1,c1})], [T_{SS}(o_{a2,c2}), T_{SR}(o_{a2,c2})]) = 0 \tag{13}$$

must be satisfied for all pairs of output values $o_{a1,c1}$ and $o_{a2,c2}$ which are stored in the same storage element s_e.

Note that these relations make no assumptions about the relative ordering of the two intervals involved, and indeed, in cases such as the operations x_3 and x_4, no information can be derived from the data flow representation to suggest an ordering. In cases where the relative order is somehow known, whether from data dependencies or external information, the above relations can be considerably simplified. For example, if operation x_3 is known to begin before operation x_4, the relation simplifies to

$$T_{XS}(x_4) \geqslant T_{XR}(x_3)$$

which simply states that execution of operation x_3 must cease before execution of operation x_4 can begin.

---

[3] An operation x_{a2} is data-dependent on an operation x_{a1} if a directed path exists in the data flow graph from an output of x_{a1} to an input of x_{a2}.

This completes the relations necessary to model a straight-line sequence of basic data transfers where no data values belong to the external environment.

E. Modeling the Interface to the External Environment

As mentioned in Section II-A external inputs and outputs can be viewed as the outputs and inputs, respectively, of a complex activity which is the external environment. They require special attention only because the external environment already exists, and the external inputs and outputs must be used and produced in a compatible manner. A value will be available (in the case of an external input) or must be made available (in the case of an external output) for some time interval $[T_1, T_2]$ given in the interface specification. This can be expressed as

$$T_1 \leqslant T_{OA}(i_{0,c}) \leqslant T_{OR}(i_{0,c}) \leqslant T_2$$

for external inputs $i_{0,c}$, and as

$$T_{IA}(o_{0,b}) \leqslant T_1 \leqslant T_2 \leqslant T_{IR}(o_{0,b})$$

for external outputs $o_{0,b}$.

F. Modeling in the Presence of Control Constructs

Control constructs will be divided into three general classes: 1) synchronization operations, 2) conditional actions, and 3) loops. This section will discuss only the timing constraints related to the control constructs. The topology of the data part is assumed to be correct. Section III-E will discuss ensuring the correct form for the data part, after some necessary variables have been introduced.

1) Synchronization Operations: Synchronization operations fall into two general categories: operations which read or write an external value, and operations which cause a transfer of control in the execution of the behavior.

The crucial point when modeling synchronization operations is that the original order of operations given in the procedural behavioral description must be preserved. The first category is included due to the (generally) unpredictable side effects of interactions with the external environment. The augmented data flow representation assumed in Section II-A includes information specifying how the operations are originally ordered. In general, the data dependencies in the data flow representation will not be sufficient to preserve this ordering. The order can be preserved by constraining outputs of the set of predominating operations in the dataflow representation to appear before the synchronization operation starts, and constraining the set of post-dominating operations in the dataflow representation to start only after the synchronization operation completes.

2) Conditional Actions: A conditional action is represented in the data flow representation as a selection between sets of values produced by the branches of the conditional action. Conditional actions can have two forms of control flow, which will be called the D-select and the C-select. Each will be discussed below.

In the D-select (data-select) form all the conditional branches are executed in parallel and the proper set of data values is selected at the end. It is the literal implementation of the data flow graph. The branches then become parallel action sequences and are modeled as outlined in the previous sections. The actual selection of a set of data values is a simple data operation and can be implemented by using a multiplexer of the appropriate size as the operator. The inputs to the operator are the value sets produced by the conditional branches, plus the selection value, and the outputs are the selected value set.

Not all conditional actions can be implemented using the D-select form. The presence, in any conditional branch, of a control construct which transfers control flow out of the branch prohibits the use of the D-select form. Array storage write operations also cannot be present in a branch unless the designer is willing to create a new copy of the array for each branch which contains a write operation.

The C-select (control select) form is the more traditional form of conditional action where the value conditioning the conditional action is used to select the branch to be executed. In fact, this value is passed to the control part, which then initiates the proper action sequence, so that the conditional action is really implemented in the control part. To model this construct properly, note that the operation of choosing a branch is a control transfer, which is the second category of synchronization operation. Prior operations must complete before the choice of a branch is made, and no operation in a branch can be initiated before the choice is made. A delay time can be assigned to the choice operation to reflect the time required for the controller to act, if the designer so desires.

Since the alternative branches of a C-select are mutually exclusive (i.e., only one is chosen for execution each time the conditional action is performed), there can be no conflicts for hardware resources between operations or data stores in different branches. Thus the hardware resource usage relations of Section II-D need not be constructed for pairs of operations or data stores when one is in one branch and the second in another. Indeed, any implementation must ensure that corresponding values in the value sets produced by the branches appear at the output of the same hardware element, so that subsequent operations may access the value without knowing the branch actually executed in the conditional action.

One must also consider the fact that, in the C-select form, certain branches of the conditional action may replace a value while others do not. In such a case, the source of the value before the conditional action must perceive the lifetime of the value as extending after the conditional action. Within the branch which modifies the value, however, the lifetime of the value ends at the point where it is replaced and the operation generating the replacement value becomes the new source.

3) Loops: When modeling a loop, two points require special consideration: the transition from one iteration to the next, and the access of values produced inside the loop by subsequent operations outside the loop. Two important assumptions are made concerning the implementation of the loop: 1) in every loop iteration, a given action utilizes the same hardware resources; 2) the initial data values fed into the loop have been placed in storage elements, and these same storage elements are used to pass values from one iteration to the next. Violation of the first assumption is equivalent to loop unwinding, a transformation beyond the scope of this modeling methodology. Two considerations motivate the second assumption. First, loops are generally expressed in this manner in procedural behavioral descriptions. Second, there must be at least one data store in every closed data path involved in the loop, in order to avoid race conditions. The simplification of the model's relations which results from placing this data store at the transition between iterations more than justifies the small restriction on the freedom of choice of implementations. These assumptions also mean that only one loop iteration needs to be modeled explicitly, which also simplifies the model.

Given these assumptions, the transition from iteration to iteration is considered first. All output values to be passed to the next iteration must be placed into the proper storage elements. This is modeled exactly as shown in Section II-C using the appropriate storage element delays. Placing all the data values transmitted from one iteration to the next into storage elements greatly simplifies the model since there is no possibility of hardware resource conflicts between iterations with this convention. Because no values other than the stored ones need remain valid across the transition, no hardware resources other than the storage elements are in use at the transition.

There are many similarities between the access of values produced inside a loop by subsequent operations outside the loop, and the access of values by operations following the C-select form of conditional action. In general, a loop may have multiple exit points. Each of these can be thought of as a branch of the C-select, producing a set of values to be passed to the subsequent operations. In addition, the lifetimes of data values with respect to their uses outside of the loop may differ from the lifetimes of data values with respect to their uses inside the loop, analogous to the manner in which the lifetime of a value may vary with respect to its usage in different branches of a C-select.

III. EXTENDING THE MODEL FOR SYNTHESIS

Section III has shown how to construct a set of algebraic relations which specify the conditions that a given RT level implementation must satisfy to have correct timing. This model can be extended for the automated synthesis of RT level data parts. The introduction of binary variables (i.e., variables limited to the values 0 and 1) into the model relations allows the inclusion of implementation decisions. These variables are used to represent the mapping of the operations and values of the data flow representation onto the operators and storage elements which compose the implementation, and to specify how operation inputs are accessed. Four additional relations using the binary variables express the conditions under which the implementation is complete and correct.

A. Representations and Definitions

In order to construct an implementation, the available hardware elements must be known. This set is given by the designer, and can consist of any hardware elements, subject only

TABLE IV
BINARY VARIABLES FOR THE SYNTHESIS MODEL

$\sigma_{d,a}$	Specifies the operation to operator mapping. $\sigma_{d,a} = 1$ indicates that operator f_d will implement operation x_a.
$\rho_{e,a,c}$	Specifies the output value to storage element mapping. $\rho_{e,a,c} = 1$ indicates that storage element s_e will be used to store output value $o_{a,c}$.
$\gamma_{a,c}$	$\gamma_{a,c} = 1$ indicates that a stored copy of output value $o_{a,c}$ exists.
$\delta_{a,b}$	Specifies how the value required by input $i_{a,b}$ is accessed. $\delta_{a,b} = 1$ indicates the stored copy of the value is accessed.
$\omega_{a,b}$	Specifies how the value required by input $i_{a,b}$ is accessed. $\omega_{a,b} = 1$ indicates the value is accessed directly from the output of the operator producing the value.

to the following restriction. Functions performed by operators must match, one-to-one, the functions of the behavioral description. However, an operator may be of arbitrary combinational complexity. (E.g., a single AND gate, a 32-bit parallel multiply, and an ALU are all acceptable.) The set should also be complete, i.e., sufficient to build at least one correct data part for the given behavioral description. This is not necessary in practice, however, as the synthesis method described in Section IV can detect when no feasible implementation exists using a given hardware set.

Let the set of operators available to implement the data part be the set

$$F = \{f_d\}$$

where f_d are instances of operator types.

It will also be necessary to refer to the subset of operators available to implement a particular data flow operation x_a; let this set be F_a. In order to ensure an optimal implementation, each set F_a should contain all elements of F capable of performing operation x_a. In practice, heuristics could be used to reduce the number of operators in each set F_a, at the risk of eliminating a possible optimal mapping of the operations of X onto the operators of F.

Let the set of storage elements available to store data flow values be the set

$$S = \{s_e\}.$$

Analogous to the set F_a for an operation, let $S_{a,c}$ be the subset of storage elements available to store an output value $o_{a,c}$. Here, also, to guarantee optimality each set $S_{a,c}$ should contain all storage elements capable of storing output $o_{a,c}$, but in practice heuristics could be used to reduce the set, with a concomitant risk of eliminating an optimal solution.

Binary variables will be used to specify three mappings, as shown in Table IV. As the algebraic relations involving these variables are developed, the reader will notice that two variables, $\gamma_{a,c}$ and $\omega_{a,b}$, are redundant. They are used in the discussion for clarity, but can be removed in programs which attempt to solve the system of relations, since the amount of computation required to solve the system depends in part on the number of binary variables in the relations.

B. Modeling Operations

In the synthesis model, a set of operators F_{a2} is assumed to be available to implement operation x_{a2}. In order for the implementation to be correct, one and only one operator should

128

be selected to implement the operation. This can be expressed by the equation

$$\sum_{d\mid f_d \in F_{a2}} \sigma_{d,a2} = 1. \tag{14}$$

It is important to realize that the binary nature of the $\sigma_{d,a}$ variables makes this summation act as a selection function. Because each $\sigma_{d,a}$ can take on only the values 0 and 1, one and only one $\sigma_{d,a}$ can be 1 if the equation is to be satisfied.

Having ensured that an operator will be assigned to execute operation x_{a2}, one must now consider the accessing of the output values $o_{a1,c1}$ which are to be supplied to the inputs $l_{a2} = \{i_{a2,b2}\}$ of the operation. The model allows two choices; a value can be accessed from the stored copy ($\delta_{a2,b2} = 1$) or directly from the output of the operator producing the value ($\omega_{a2,b2} = 1$).

Again, one and only one access method must be selected, and this can be expressed with the equation

$$\omega_{a2,b2} + \delta_{a2,b2} = 1 \tag{15}$$

for each input $i_{a2,b2}d_{la2}$. Here, too, the binary nature of the variables is crucial to the correctness of the equation.

In addition, a value certainly cannot be accessed from a stored copy if no such copy exists; hence it must also be true that

$$\delta_{a2,b2} \leqslant \gamma_{a1,c1} \tag{16}$$

for each $i_{a2,b2} \in l_{a2}$. The relation assumes $o_{a1,c1}$ is the value to be supplied to input $i_{a2,b2}$.

Now an expression for $T_{IA}(i_{a2,b2})$ must be defined, keeping in mind that a priori knowledge of where the input $i_{a2,b2}$ will access its source value $o_{a1,c1}$ is no longer available. The solution is to combine the two equations (1) and (2) of Section II-B into the single equation

$$(1),(2): \quad T_{IA}(i_{a2,b2}) = \omega_{a2,b2} T_{OA}(O_{a1}) + \delta_{a2,b2} T_{SA}(o_{a1,c1}) \tag{17}$$

thus taking into account the fact that when a value can be accessed will depend on how it is accessed. (The reference numbers to the left of the extended relations in this section refer to the original forms in the previous section.) To understand this equation, recall that equation (15) states that one and only one of the variables $\omega_{a2,b2}$ and $\delta_{a2,b2}$ can have the value 1; the other must be 0. Thus, the proper way to interpret the equation is that $\omega_{a2,b2}$ and $\delta_{a2,b2}$ select the proper value, either $T_{OA}(O_{a1})$ or $T_{SA}(o_{a1,c1})$, respectively. For example, if $\omega_{a2,b2} = 1$, indicating that the value $o_{a1,c1}$ is to be accessed from the output of the operator which is producing it, then $T_{OA}(O_{a1})$, the time the value becomes available at the operator output, will be the time selected as $T_{IA}(i_{a2,b2})$, the time the value is available for use by the input. There is really no addition involved, as one of the terms is always zero.

The relation governing the time an operator can begin to perform an operation, $T_{XS}(x_a)$, is unchanged. One must still have

$$(3): \quad T_{XS}(x_a) \geqslant \max_{b\mid i_{a,b} \in l_a} T_{IA}(i_{a,b}). \tag{18}$$

The expression for the time $T_{OA}(O_a)$ when the outputs O_a of operation x_a will be available must also allow for uncertainty. Any operator from the set F_a could be selected to implement the operation. Here again, it is possible to take advantage of the fact that one and only one of the mapping variables $\sigma_{d,a}$ can be 1 and write

$$(4): \quad T_{OA}(O_a) = T_{XS}(x_a) + \sum_{d\mid f_d \in F_a} \sigma_{d,a} D_{FP}(f_d). \tag{19}$$

There is really no summation performed, but rather a selection, since all but one of the terms are zero. The nonzero term is the propagation delay of the operator selected to perform the operation.

This completes the synthesis model for the forward aspect of an operation. The relations for the backward aspect will be developed next.

In Section II-B, two relations were given to define the time $T_{OR}(o_{a1,c1})$ for which an output value must remain valid. One, (5), applies to the case where the value is stored, and the other, (6), applies to the case where the value is used directly. In the synthesis model, it is not known which case will be required; thus, the relations must be conditional on the occurrence of the case they model. This can be achieved with the relations

$$(5): \quad T_{OR}(o_{a1,c1})$$
$$\geqslant \gamma_{a1,c1}\left[T_{SS}(o_{a1,c1}) + \sum_{e\mid s_e \in S_{a1,c1}} \rho_{e,a1,c1} D_{SH}(s_e) \right] \tag{20}$$

and

$$(6): \quad T_{OR}(o_{a1,c1}) \geqslant \max_{(a2,b2)\mid o_{a1,c1}=\mathrm{src}(i_{a2,b2})} \omega_{a2,b2} T_{IR}(l_{a2}) \tag{21}$$

where $o_{a1,c1} = \mathrm{src}(i_{a2,b2})$ indicates that the value $o_{a1,c1}$ is required by the input $i_{a2,b2}$. The reader will recognize the summation as being in fact a selection of the hold time for the storage element s_e chosen to store output value $o_{a1,c1}$. (Relation (25) will ensure that at most one variable $\rho_{e,a1,c1}$ takes on the value 1.) Recalling that $\gamma_{a1,c1}$ indicates whether value $o_{a1,c1}$ is stored, it can be seen that in the case where $o_{a1,c1}$ is not stored and $\gamma_{a1,c1} = 0$, the relation (20) reduces to the trival condition

$$T_{OR}(o_{a1,c1}) \geqslant 0$$

which must certainly be true in any real system. Thus when output value $o_{a1,c1}$ is not stored, the relation is trivially satisfied; this is the conditional property we desired. Similarly, if output value $o_{a1,c1}$ is never accessed directly, the inequality (21) is trivially satisfied.

Proceeding backward through the operation, the relation for $T_{XR}(x_a)$, the time an operation can cease execution, is

$$(7): \quad T_{XR}(x_a) \geqslant \max_{(a,c)\mid o_{a,c} \in O_a} T_{OR}(o_{a,c}) \tag{22}$$

and the equation for $T_{IR}(l_a)$, the time the operation **input** values are no longer required, is

$$(8): \quad T_{IR}(l_a) = T_{XR}(x_a) - (1 - \chi) \sum_{d | f_d \in F_a} \sigma_{d,a} D_{FP}(f_d),$$

$$0 \le \chi \le 1. \quad (23)$$

These expressions complete the set of relations required to describe an operation in this synthesis model.

C. Modeling the Store of a Value

Again, as with the model of Section III, an expression must be constructed which states that a value cannot be stored until it has been present at the storage element input for at least the input setup time. Now, however, it is not known which storage element will be used, and one must also consider the possibility that the value will not be stored at all. A relation which allows for these uncertainties is

$$(9): \quad T_{SS}(o_{a,c}) \ge \gamma_{a,c} \left[T_{OA}(O_a) + \sum_{e | s_e \in S_{a,c}} \rho_{e,a,c} D_{SS}(s_e) \right] \quad (24)$$

which should be interpreted in much the same manner as relation (20).

The defining equation for $\gamma_{a,c}$ is

$$\gamma_{a,c} \ge \sum_{e | s_e \in S_{a,c}} \rho_{e,a,c}. \quad (25)$$

Note that since $\gamma_{a,c}$ is binary, at most one variable $\rho_{e,a,c}$ can take the value 1, and thus at most one storage element can be selected to store the value $o_{a,c}$. Equation (25) does not require that at least one $\rho_{e,a,c}$ be 1 since storing an output value $o_{a,c}$ is an optimal action, and if the value is not stored, no storage element need be assigned.

The equation defining the variable $T_{SA}(o_{a,c})$, the time a stored value becomes available at the storage element output, also uses the "selection by summation" construct, and it becomes

$$(10): \quad T_{SA}(o_{a,c}) = T_{SS}(o_{a,c}) + \sum_{e | s_e \in S_{a,c}} \rho_{e,a,c} D_{SP}(s_e). \quad (26)$$

Finally, an expression is needed for $T_{SR}(o_{a1,c1})$, the lifetime of the value $o_{a1,c1}$ in the storage element. This relation is

$$(11): \quad T_{SR}(o_{a1,c1}) \ge \max_{(a2,b2) | o_{a1,c1} = \text{src}(i_{a2,b2})} \delta_{a2,b2} T_{IR}(l_{a2}). \quad (27)$$

It is the analog, for the stored copy of the output value, of relation (21) for operation outputs.

This completes the set of relations used in the synthesis model to describe the store of a value.

D. Modeling Hardware Resource Sharing

Despite the fact that the hardware elements which will actually be used in an implementation, and the times they will be used, are not known precisely, it is still necessary to prevent overlaps in the scheduling of events on hardware elements. The satisfaction of the relations

$$(12): \quad \sum_{a1,a2 | f_d \in F_{a1} \wedge f_d \in F_{a2}} \sigma_{d,a1} \sigma_{d,a2} L([T_{XS}(x_{a1}),$$

$$T_{XR}(x_{a1})], [T_{XS}(x_{a2}), T_{XR}(x_{a2})]) = 0 \quad (28)$$

for operators and

$$(13): \quad \sum_{(a1,c1),(a2,c2) | s_e \in S_{a1,c1} \wedge s_e \in S_{a2,c2}}$$

$$\cdot \rho_{e,a1,c1} \rho_{e,a2,c2} L([T_{SS}(o_{a1,c1}), T_{SR}(o_{a1,c1})],$$

$$[T_{SS}(o_{a2,c2}), T_{SR}(o_{a2,c2})]) = 0 \quad (29)$$

for storage elements will ensure that no physical hardware element is assigned to do two things at once. Using equation (28) as an example, the summation is taken over all unique pairs of operations x_{a1} and x_{a2} such that the sets of operators F_{a1} and F_{a2} available to implement the operations contain the common operator f_d. The summation here is really a summation, rather than a selection as it was used in the previous relations. (However, since both the ρ and σ variables and the overlap function L can take on only binary values, the relations could be expressed in Boolean sum-of-products form. The important issue is simply to determine if any term is nonzero.) Note that an individual term can become zero in two ways.

1) If operator f_d is not actually chosen to implement both x_{a1} and x_{a2}, the product $\sigma_{d,a1} \sigma_{d,a2}$ will be zero. The overlap portion of the term is essentially ignored, as it should be, since the operations are being performed on different operators.

2) If both $\sigma_{d,a1}$ and $\sigma_{d,a2}$ are 1, both operations are using operator f_d and the overlap function must be zero to satisfy the equation.

As noted in Section II-D these relations can be simplified if there is *a priori* knowledge of the relative orders of the two events. For example, for the storage of a single value, equation (29) simplifies to

$$T_{SS}(o_{a2,c2}) \ge \rho_{e,a1,c1} \rho_{e,a2,c2} T_{SR}(o_{a1,c1})$$

if $T_{SS}(o_{a1,c1})$ is known to precede $T_{SS}(o_{a2,c2})$. (I.e., if the storage of the value $o_{a1,c1}$ is known to preceed the storage of the value $o_{a2,c2}$.)

This completes the set of relations required by the synthesis model to describe a straight-line sequence of data transfers. The complete set is summarized in Table V. The number to the left of each equation is the reference number of the equation in the text.

E. Modeling in the Presence of Control Constructs

Section II-F dealt with the timing relations for control constructs, and alluded to the necessity of knowing the detailed sequence of operations evoked by the controller in order to determine if a data part implementation is correct. In particular, knowledge is needed about where and when assignment to storage elements is performed. The introduction of the variables $\rho_{e,a,c}$ and $\delta_{a,b}$ allow the expression of conditions on data storage, and this section deals primarily with this issue. Of course, the timing considerations of Section II-F remain valid. No additional relations are required for synchronization operations, and they are not discussed here.

130

TABLE V
SUMMARY OF RELATIONS FOR SYNTHESIS MODEL

(14): $\sum_{d|f_a \in F_a} \sigma_{d,a} = 1$

(15): $\omega_{a2,b2} + \delta_{a2,b2} = 1$

(17): $\delta_{a2,b2} \leq \gamma_{a1,c1}$

(17): $T_{IA}(i_{a2,b2}) = \omega_{a2,b2} T_{OA}(O_{a1}) + \delta_{a2,b2} T_{SA}(o_{a1,c1})$

(18): $T_{XS}(x_a) \geq \underset{b|i_{a,b} \in I_a}{MAX}\, T_{IA}(i_{a,b})$

(19): $T_{OA}(O_a) = T_{XS}(x_a) + \sum_{d|f_a \in F_a} \sigma_{d,a} D_{FP}(f_d)$

(20): $T_{OR}(o_{a1,c1}) \geq \gamma_{a1,c1}\left[T_{SS}(o_{a1,c1}) + \sum_{e|s_e \in S_{a1,c1}} \rho_{e,a1,c1} D_{SH}(s_e)\right]$

(21): $T_{OR}(o_{a1,c1}) \geq \underset{(a2,b2)|o_{a1,c1} = src(i_{a2,b2})}{MAX}\, \omega_{a2,b2} T_{IR}(i_{a2,b2})$

(22): $T_{XR}(x_a) \geq \underset{(a,c)|o_{a,c} \in O_a}{MAX}\, T_{OR}(o_{a,c})$

(23): $T_{IR}(I_a) = T_{XR}(x_a) \cdot (1-\chi) \sum_{d|f_a \in F_a} \sigma_{d,a} D_{FP}(f_d), \qquad 0 \leq \chi \leq 1$

(24): $T_{SS}(o_{a,c}) \geq \gamma_{a,c}\left[T_{OA}(O_a) + \sum_{e|s_e \in S_{a,c}} \rho_{e,a,c} D_{SS}(s_e)\right]$

(25): $\gamma_{a,c} \geq \sum_{e|s_e \in S_{a,c}} \rho_{e,a,c}$

(26): $T_{SA}(o_{a,c}) = T_{SS}(o_{a,c}) + \sum_{e|s_e \in S_{a,c}} \rho_{e,a,c} D_{SP}(s_e)$

(27): $T_{SR}(o_{a1,c1}) \geq \underset{(a2,b2)|o_{a1,c1} = src(i_{a2,b2})}{MAX}\, \delta_{a2,b2} T_{IR}(i_{a2})$

(28): $\sum_{a1,a2|f_{a1} \in F_{a1} \wedge f_a \in F_{a2}} \sigma_{d,a1}\sigma_{d,a2} L\left([T_{XS}(x_{a1}),T_{XR}(x_{a1})]\cdot[T_{XS}(x_{a2}),T_{XR}(x_{a2})]\right) = 0$

(29): $\sum_{(a1,c1),(a2,c2)|s_e \in S_{a1,c1} \wedge s_e \in S_{a2,c2}} \rho_{e,a1,c1}\rho_{e,a2,c2} L\left([T_{SS}(o_{a1,c1}),T_{SR}(o_{a1,c1})]\cdot[T_{SS}(o_{a2,c2}),T_{SR}(o_{a2,c2})]\right) = 0$

1) Conditional Actions: For conditional actions, the need is to be able to specify the common hardware access point for the alternative sources of each value produced by the conditional action. If a common access point were not provided for each of the values, it would be necessary for the succeeding operations to know which branch was executed in order to access the proper source of the value. This would have the effect of extending the unique control sequences for each branch until the last use of the values.

One way to provide a common access point is to use storage elements. (An alternative is to specify that a value will appear on a particular data path, but this model does not allow the specification of an individual data path.) One storage element is provided for each value produced by the conditional action, and all branches use that storage element for the value. Subsequent operations then access these stored copies. To insure that each alternative source of a value uses the same storage element, one simply uses the same set of storage elements $S_{a,c}$ and variables $\rho_{e,a,c}$ in the relations dealing with the storage of each alternative, forcing the store to occur in all cases.

Along the same lines as the above method, it will sometimes be possible to assign operators to operations in such a way that common access points are maintained. This is done by assigning the same operator set F_a and variables $\sigma_{d,a}$ to multiple operations. Of course, the types of operations involved and the types of operators available will sometimes preclude this option, since it must be possible to implement all operations which produce a given value with the same operator.

Finally, as in the D-select implementation, a multiplexer or

other gating operator may be used to provide the common access point. It would be modeled in the same manner as suggested for the D-select.

2) Loops: As with conditional actions, loops require common access points for alternative sources of values. Here there are two sets of values to consider: 1) the set of values brought into the loop and passed from iteration to iteration; and the set of values passed out of the loop.

The first set of values has at least two sources; more are possible if the loop contains multiple RESTART operations. The number of alternative sources for the second set of values depends on the number of LEAVE operations in the loop. The solution in the first case is the common register method of the previous section, given the convention for storage at the iteration boundary adopted in Section II-F-3. In the second case, both the common register and common operator methods could be used. The use of a multiplexer could only be attempted if all of the LEAVE operations resided in branches of a single conditional action within the loop, or if some new value could be derived to condition the multiplexer.

F. Estimating the Size of the Model

The reader has probably noticed the apparent conflict between the nonlinear form of some of the relations in Table V and the stated intention of solving the model as a mixed-integer linear programming problem. The relations are linearized by splitting some, and/or by adding additional binary variables. The nonlinear forms have been used in this paper as the linearization tends to obscure the physical interpretations of the relations. The estimates of model size which follow, however, take into account the increase in relations and binary variables due to linearization.

Let n be a composite estimate of the number of operations and values in the data flow description, and let h be a composite estimate of the number of hardware elements available for the implementation. Then the following statements can be made about the worst case growth of the synthesis model.

The number of algebraic relations grows as $O(hn^2)$. This bound is due to the number of hardware usage overlap equations (28) and (29). All other types of relations grow linearly with respect to the size of the data flow description, i.e., $O(n)$.

The number of continuous (timing) variables grows as $O(n)$.

The number of binary (synthesis) variables grows as $O(hn^2)$. This bound is due to the introduction of one binary variable to linearize each hardware usage overlap equation. All other types of binary variables grow as $O(hn)$.

IV. A SYNTHESIS EXAMPLE

The synthesis model has been tested by using it to generate data part implementations for small, simple behavioral descriptions. In this section the results for one such description will be discussed in detail, along with some general observations from results for other descriptions.

The source language for the behavioral descriptions is ISPS [2]. "Small" is used to mean five to seven RT level operations; "simple" excludes control operations such as those discussed in Section II-F. The ISPS description is used as input

131

```
Criss.Cross :=

BEGIN

   ** Carriers **
   ti<0:15> ,
   t2<0:15> ,
    a<0:15> ,
    b<0:15> ,

   ** Activity **
   action  :=
   ( t1 ← a+b next
     t2 ← a-b next
     a ← t1+t2 next
     b ← t1-t2 )

END
```

(a)

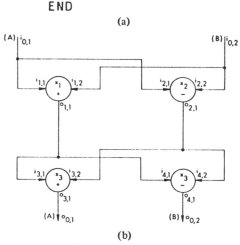

(b)

Fig. 5. (a) Criss-cross ISPS behavioral description. (b) Criss-cross data flow representation.

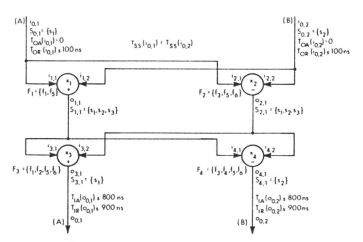

Fig. 6. Restrictions on the implementation of criss cross.

TABLE VI
HARDWARE ELEMENTS AVAILABLE FOR CRISS-CROSS IMPLEMENTATION

storage	bits	IC's	D_{SS}	D_{SH}	D_{SP}	cost
s_1	<16>	(2)SN74273	20ns.	0 ns.	27 ns.	\$8.10
s_2	<16>	(2)SN74273	20ns.	0 ns.	27 ns.	\$8.10
s_3	<16>	(2)SN74273	20ns.	0 n..	27 ns.	\$8.10

operator	bits	IC's	function	D_{FP}	cost
f_1	<16>	(4)SN7483	+	70 ns.	\$14.20
f_2	<16>	(4)SN7483	+	70 ns.	\$14.20
f_3	<16>	(2)81LS96	-	85 ns.	\$24.18
		(4)SN7483			
f_4	<16>	(2)81LS96	-	85 ns.	\$24.18
		(4)SN7483			
f_5	<16>	(4)SN74181	ALU	107 ns.	\$19.00
f_6	<16>	(4)SN74181	ALU	107 ns.	\$19.00

to a translator which generates the data flow version of the behavioral description. The Value Trace (VT) form [15], [19] is used for the data flow specification. The VT specification, a description of the set of hardware elements available, a specification of the interface between the design and the outside world, and cost and performance constraints provide the input to the program which generates the system of equations and inequalities which comprise the model. This program also gives the designer a means to specify an objective function to be used to select the optimal design. The model, now viewed as a system of constraints, is solved as a mixed-integer linear programming problem to maximize the objective function (i.e. minimize cost). This is done with the BANDBX program [12], which uses a branch and bound technique to solve the problem. The restriction to small descriptions is due to limitations imposed by BANDBX; the restriction to simple descriptions due to limitations of the current version of the model generator program. Both are practical limits of the software available at this time.

Using the procedure described in the previous paragraph, a set of implementations was generated for the behavioral description of Fig. 5. The VT representation has 4 RT level operations (2 additions and 2 subtractions) and 6 values for which storage can be allocated. Each implementation was optimized for lowest cost, where cost was calculated as the sum of the costs of the operators and registers used in the implementation. By steadily increasing the performance required of an implementation while retaining the least-cost objective, the noninferior surface of the design space of implementations with respect to cost and performance was outlined. Several additional constraints were placed on the implementation, as shown in Fig. 6. The period during which the external inputs l_0 are valid has been restricted to the interval 0 to 100 ns., and a single control signal is assumed to latch the inputs in registers, if they are stored. (This is expressed by equating the storage times $T_{SS}(i_{0,1})$ and $T_{SS}(i_{0,2})$.) The permitted mappings of operations and values to hardware elements are indicated by the sets of hardware elements listed beside the data flow elements in the picture. (For example, the operation x_3 can be implemented by any of the operators f_1, f_2, f_5, or f_6, but the value $i_{0,1}$ can only be stored in register s_1.) Although not shown in Fig. 5-2, the output values $o_{3,1}$ and $o_{4,1}$ were forced to be stored and the external outputs $o_{0,1}$ and $o_{0,2}$ forced to access the stored copies by setting the variables $\rho_{1,3,1}, \rho_{2,4,1}, \delta_{0,1}$, and $\delta_{0,2}$ to 1.

The hardware elements made available for the implementation, along with costs and timing information, are shown in Table VI.

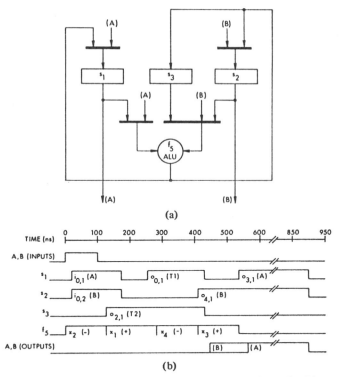

(a)

(b)

Fig. 7. (a) Implementation 1 for criss cross. (b) Timing for (a).

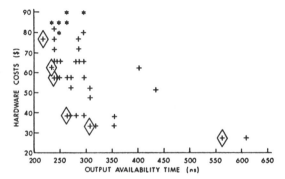

Fig. 8. Design space of implementations for criss cross.

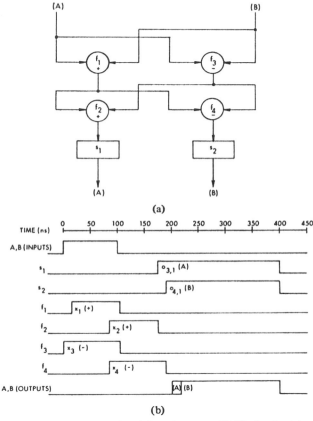

(a)

(b)

Fig. 9. (a) Implementation 6 for criss cross. (b) Timing for (a).

The model generated for this problem contains 124 relations in 87 variables. Of the relations, 60 are used to enforce the behavior specification, 16 to ensure the implementation is physically correct, 38 to ensure that the hardware elements are properly used, and 10 for linearization of the cost objective. Of the variables, 41 are continuous variables as described in Table II and provide timing information. Of the 46 integer variables, 13 specify the mapping of operations to operators, 8 the mapping of values to registers, 8 the accessing of values by operation inputs, and 17 are necessary for constraint linearization.

For the first implementation, it was assumed that the outputs had to be valid at the time $t = 800$ ns. and remain valid until the time $t = 900$ ns. These are the restrictions on the outputs shown in Fig. 6. The optimal implementation is shown in Fig. 7(a), and the timing diagram showing the usage of the hardware is shown in Fig. 7(b). The ALU f_5 is used for all operations, and all three registers are required for storage of the inputs, intermediate results, and outputs. The last output value produced, $A(o_{0,1})$, becomes valid at the time $t = 562$ ns. Obviously, there are several other permutations of the hardware and data paths which are equivalent in cost and performance.

In order to see the tradeoffs made as performance requirements were increased, the time at which the values of A and B were required to be available was decreased to just under the times for the previous solution and the model was solved again. Repetition of this process until no feasible implementation could be found produced the cost/performance curve of Fig. 8. The fastest implementation is shown in Fig. 9. In Fig. 8 a diamond indicates an optimal design; a plus or asterisk indicates a sub-optimal design produced by BANDBX while search-

ing for an optimal solution. At all times, the cheapest design meeting the performance requirements was produced.

The designs synthesized by the system have been correct within the accuracy of the model for all examples thus far. Hardware elements are well utilized, with a minimum of idle time. It is apparent, though, that more consideration must be given to interconnections and switching. Permutations of a set of hardware elements which are equivalent in cost and performance under this model can display considerable differences when interconnections are taken into account.

The best designs at each performance level were all found in under 15 min of CPU time, reflecting the tendency of the branch and bound technique to find optimal or near optimal solutions quickly, spending the majority of its time verifying optimality. In this example, only the design of Fig. 7 is verified to be optimal, with 33 min of CPU time required to exhaustively search the branch and bound tree. In all other cases, the tree was not exhaustively searched within 1 h of CPU time.

V. Conclusions

This paper has presented a modeling methodology whereby the essential properties necessary to ensure correct hardware implementations of a given behavior can be formally expressed as a system of algebraic relations. The adequacy of the methodology has been verified by using it to create a model suitable for the synthesis of RT level data parts for digital systems, and by producing correct, optimal data part designs by solving the model as a MILP problem.

Two qualities of the model are: 1) models can be created at many levels of detail, depending on the particular features of a digital system that are to be emphasized; and 2) the variables and relations of the model are related to physical events in, and properties of, a hardware implementation in a direct, natural manner.

With respect to data path synthesis at the RT level, 1) the methodology provides a single, unified means of expressing the decisions and tradeoffs involved in synthesizing a hardware implementation from a given behavioral description, 2) the algebraic relations clearly and explicitly show the complicated interrelations between the form of the data part and the order in which the behavioral actions are executed, and 3) the designer can easily specify particular characteristics which, for arbitrary reasons, must exist in the data part.

In evaluating the effectiveness of mixed-integer linear programming as a solution technique, however, it is impossible to avoid the conclusion that, if MILP is to be used, it must be efficiently implemented and highly specialized for precisely this task. Efficient data structures and algorithms for large, sparse systems must be introduced as part of the solution procedure. They were not present in the BANDBX program, and this is reflected by the very large solution times reported here, even for small examples. The addition of these specialized techniques promises significant improvement in solution time.

Other areas to which a model of this type might be applied include design verification at the RT level, and engineering changes to existing systems. Both of these applications avoid the computational difficulties which hamper the application of this modeling methodology for synthesis.

Acknowledgment

The authors would like to acknowledge the generosity of M. McFarland, who developed the Value Trace software, and E. Balas, who provided access to the BANDBX program. Thanks are due also to Jon Bentley, Steve Director, and Sarosh Talukdar for their guidance, and to the Design Research Center at Carnegie–Mellon University for providing the multidisciplinary environment which nutured this work. The authors would also like to acknowledge the thoughtful comments of the referees, which (we hope) have improved the clarity of the presentation. Finally, we would like to thank the CMU Computer Science Department for unstintingly providing computing resources.

References

[1] M. Barbacci and D. Siewiorek, "The CMU RT-CAD system: An innovative approach to computer aided design," in *Amer. Fed. Info. Processing Soc. Conf. Proc.*, vol. 45, June 1976, pp. 643–655.

[2] M. Barbacci, G. Barnes, R. Cattell, and D. Siewiorek, "The symbolic manipulation of computer descriptions; The ISPS computer description language, Dept. Computer Sci., Carnegie–Mellon Univ., Pittsburgh, PA, Tech. Rep., Aug. 1979.

[3] W. Chu, L. Holloway, M. Lan, and K. Efe, "Task allocation in distributed data processing," *Computer*, vol. 13, no. 11, pp. 57–69, Nov. 1980.

[4] M. Ciesielski and E. Kinnen, "An optimum layer assignment for routing in ICs and PCBs," in *Design Automation Conf. Proc.*, ACM SIGDA, IEEE Computer Society-DATC, 1981, no. 18, pp. 733–737.

[5] J. Darringer and W. Joyner, "A new look at logic synthesis," in *17th Design Automation Conf. Proc.*, ACM SIGDA, IEEE Computer Society-DATC, June 1980, pp. 543–549.

[6] I. Grossman and J. Santibanez, "Application of mixed-integer linear programming in process synthesis," Design Research Center, Pittsburgh, PA, Tech. Rep. DRC-06-12-79, Sept. 1979; also, to be published in *Computers and Chemical Engineering*.

[7] G. Hachtel, "On the sparse approach to optimal layout," in *IEEE Int. Conf. Circuits Computers Proc.*, N. Rabbat, Ed., IEEE Circuits and Systems Society, IEEE Computer Society, Oct. 1980, pp. 1019–1022.

[8] L. Hafer and A. Parker, "Automated synthesis of digital hardware," *IEEE Trans. Computers*, vol. C-31, no. 2, pp. 93–109, Feb. 1981.

[9] L. Hafer, "Automated data-memory synthesis: A formal model for the specification, analysis and design of register-transfer level digital logic," Ph.D. thesis, Dep. Elec. Eng., Carnegie–Mellon Univ., Pittsburgh, PA, May 1981.

[10] D. Hinshaw and K. Irani, "Optimal selection of functional components for microprogrammable central processing units," in *Fifth Annu. Workshop Microprogramming Proc.*, IEEE Computer Society, ACM SIGMICRO, Sept. 1971, pp. 8–27.

[11] G. Leive, *The SYNNER's Guide*. Pittsburgh, PA: Carnegie-Mellon University, 1980.

[12] C. Martin, *BANDBX: An Enumeration Code for Pure and Mixed Zero-One Programming Problems*. Columbus, OH: Ohio State University, 1978.

[13] P. Marwedel, "The MIMOLA design system: Detailed description of the software system," in *16th Design Automation Conf. Proc.*, ACM SIGDA, IEEE Computer Society-DATC, June 1979, pp. 50–63.

[14] G. McClain, "Optimal Design of Central Processor Data Paths," Ph.D. thesis, Dep. Elec. Computer Eng., Univ. Michigan, Ann Arbor, MI, Apr. 1972.

[15] M. McFarland, "The value trace: A data base for automated digital design," Master's thesis, Dep. Elec. Eng., Carnegie–Mellon Univ., Pittsburgh, PA, Dec. 1978.

[16] ——, "On proving the correctness of optimizing transformations in a digital design automation system," in *Design Automation Conf. Proc.*, ACM SIGDA, IEEE Computer Society-DATC, no. 18, June 1981, pp. 90–97.

[17] A. Nagle and A. Parker, "Algorithms for multiple-criterion design of microprogrammed control hardware," in *Design Automation Conf. Proc.*, ACM SIGDA, IEEE Computer Society-DATC, no. 18, June 1981, pp. 486–493.

[18] S. Papoulias and I. Grossmann, "Optimal synthesis of integrated utility systems," in *Amer. Inst. Chem. Eng. Meet. Proc.*, Apr. 1981.

[19] E. Snow, D. Siewiorek, and D. Thomas, "A technology-relative computer aided design system: Abstract representations, transformations, and design tradeoffs," in *Design Automation Conf. Proc.*, ACM SIGDA, IEEE Computer Society DATC, no. 18, June 1978, pp. 220–226.

[20] D. Thomas and D. Siewiorek, "Measuring designer performance to verify design automation systems," in *Design Automation Conf. Proc.*, ACM SIGDA, IEEE Computer Society-DATC, no. 14, June 1977.

[21] G. Zimmermann, "The MIMOLA design system: A computer aided digital processor design method," in *Proc. 16th Design Automation Conf.*, ACM SIGDA, IEEE Computer Society-DATC, June 1979, pp. 53–58.

Author Index

A

Agarwal, V. K., 75

B

Bartlett, K., 34
Berman, C. L., 14
Brayton, R. K., 48

C

Cain, R. G., 54
Cohen, W., 23, 34

D

Dagenais, M. R., 75
Darringer, J. A., 14
de Geus, A. J., 23, 34
De Micheli, G., 84

H

Hachtel, G., 34
Hafer, L. J., 121
Hong, S. J., 54

J

Joyner, W. H., Jr., 14

M

McMullen, C., 48

N

Newton, A. R., 3

O

Ostapko, D. L., 54

P

Parker, A. C., 121

R

Rudell, R. L., 70
Rumin, N. C., 75

S

Sangiovanni-Vincentelli, A. L. M., 3, 70
Siewiorek, D. P., 104

T

Trevillyan, L., 14
Tseng, C-J., 104

Editor's Biography

A. Richard Newton (Member, IEEE) was born in Melbourne, Australia, on July 1, 1951. He received the B.Eng.(elec.) and M.Eng.Sci. degrees from the University of Melbourne, Melbourne, Australia in 1973 and 1975, respectively, and the Ph.D. degree from the University of California, Berkeley, in 1978.

He is currently a Professor and Vice Chairman in the Department of Electrical Engineering and Computer Sciences, University of California, Berkeley and a consultant to a number of companies for computer-aided design of integrated circuits. His research interests include all aspects of the computer-aided design of integrated circuits with emphasis on simulation, automated layout techniques, and design methods for VLSI integrated circuits.

Dr. Newton is a member of Sigma Xi.